The Automotive Industry
and the
Global Environment

The Next 100 Years

zle

e

gineers, Inc.
Warrendale, Pa.

Copyright © 1999 Ford Motor Company
World Headquarters Building, Suite 740
American Road
Dearborn, Michigan 48124

ISBN 0-7680-0439-X ✓

SAE Order No. R-263

Table of Contents

Abstract

The first century of the automobile has had a profound effect on society, the economy, industry, technology, and the global environment. One of the great challenges of the past century has been to minimize the potential impact of transportation systems on the global environment. The challenges of the next 100 years will be much greater than those of the first century, especially for the global, urban-centered economies of the next century. During this next era, automotive companies will not only need to be concerned with the effect of transportation systems on the environment and natural resources but how sociological, cultural, economic and political factors may affect these systems. Some key imperatives of the 21st century will be to advocate global approaches to the establishment of international standards; champion consortia in cooperation with government, academia and industry; develop new approaches that integrate technical, social and economic solutions; balance free-market with regulatory approaches; support scientific research as needed to help develop sound environmental policies; and use life-cycle assessment models to help evaluate the potential environmental, economic and energy outcomes of transportation systems on a global basis. In this paper we focus on many of these issues, especially as they relate to developing countries such as China and their impact on energy resources and the environment. As the global resources of petroleum become depleted, there will be a gradual shift to other alternative fuel and energy sources. The use of electric power for transportation applications will become more attractive since this electricity can be generated from a wide array of sources, each of which can be easily feed into the world's extensive electrical distribution systems. In order to effectively utilize emerging energy sources, global transportation companies will continue to accelerate the development of cost-effective and reliable technologies including alternative fuel vehicles, high-energy efficiency vehicles, small personal-use vehicles, electricycles and intelligent transportation systems. Personal transportation systems will become highly integrated with public transportation strategies. The energy efficient vehicles of the future will incorporate "hybrid-electric propulsion systems." Such systems may employ various combinations of technologies including: a solid-state energy storage system which may be electrochemical or thermal; a fuel tank to store liquid or gaseous fuels; an internal combustion engine, heat engine or fuel cell stack; and a direct-drive electric motor. As has been the case in the 20th century, the ultimate success of these new technologies in the 21st century will be based primarily upon customer acceptance which includes value, functionality and reliability.

Introduction

Evolution from the Industrial to the
Global Revolution

The 18th century was born amidst an explosion of human productivity referred to as the Industrial Revolution. During this period, which accelerated dramatically in the Nineteenth and Twentieth centuries, new inventions made it possible for industrial production to evolve from small cottage industries to larger factories. Mechanized production and assembly lines made it practical for workers to become more efficient, the prices of produced goods to drop, and formerly scarce items to become affordable to a much larger proportion of the population.

The steam engine, invented during the 1830's in England, was one of the most important developments of the Industrial Revolution. The steam engine provided a dramatic improvement in the way goods were manufactured and it also provided more efficient and faster modes of transportation of goods and people by train and ships. Engines also made the automobile and the truck possible, signaling the beginning of a new era. The internal combustion engine, together with oil and gasoline fuels, which were also produced by mass production methods, made it possible for individuals and small groups of people to travel with unprecedented freedom. Industries grew up and increasingly utilized the motorized truck to replace horses.

During the "First Revolution" of the automobile, scientists, engineers, science fiction writers and the entertainment industry provided the world with many future visions of vehicles that are intelligent and able to carry on intellectual discussions; can operate in auto-pilot corridors on the ground or in the air; can accurately monitor their location and the location of other objects near the vehicle; could utilize a variety of energy resources and have little or no effect on the global environment.

Some early predictions have come true such as the formation of suburbs around cities, while others such as rapid transit (trolley car) lines to all suburban areas (Walter, 1992) and auto-pilot corridors in the urban areas have not yet been fulfilled. The primary problem with these early predictions was that a total systems analysis was not carried out to determine the practicality of implementing such technologies. We now appreciate that there are many factors that can influence the acceptance of new technologies including cost, safety, environmental and energy requirements, convenience, infrastructure demands, social needs, government requirements, demographics and many others.

4

As we enter the next phase of the Industrial Revolution, which we believe will be a "Global Revolution", we may ask how accurate will past and current visionaries be in their predicitions? How can the automobile industry meet the environmental and energy challenges of the future? The primary objective of this paper is to crack open the door of the future to help provide the reader with a glimpse of the exciting possibilities for automotive industry and the travelers of the 21st century.

At the beginning of this century, there was much experimentation. We are now entering a new century in which there will be a great period of change because of many new driving forces, two of which will be energy and the environment. Although scientists and engineers are working on the development of many new technologies, most of them have a low probability of becoming successful. However, a major success in any one of these technologies could change the future direction of all others.

Historians often say that the lessons of the past can help provide a vision of the future and therefore reduce our chances of traveling down the wrong road. Chapter 1 briefly describes the relationship between the "Industrial Revolution", the development of the automotive industry, and the important role that technology has played in the development of this industry.

Chapters 2 and 3 present a perspective and prospective analysis on the potential effects of globalization, the environment and energy challenges, market economy growth, and population dynamics on the future automotive industry. The environmental and energy challenges developed from this analysis are used to formulate several approaches and methodologies recommended in Chapter 4.

Chapter 5 provides a future prospective on the second revolution of the automotive industry and the most likely scenarios for transportation technologies of the 21st century. These new transportation systems will need to overcome future challenges by meeting the ever growing global demand for transportation goods and services while minimizing adverse energy, environmental and economic impacts. It is postulated that the successful implementation of these technologies will be based primary on value to the customer in terms of cost, quality and the ability of the product to meet their needs.

This paper was not meant to be exhaustive or provide an impression that the authors are possessed with clairvoyance. Each of the subjects included in this paper are covered in more detail by cited publications. However, to our knowledge this is the first attempt by anyone to integrate and discuss most of the potential issues related to the automobile industry and the global environment - past, present and future.

Chapter 1
The Industrial Revolution
and the Automobile

It was more than a century ago, when the world's first automobile, the Benz Patent Motorwagen changed the course of transportation forever. The revolutionary three-wheeler built by Carl Benz in 1885 ignited the spark that started the auto industry. Henry Ford completed his first car, the Quadricycle, and took it for a drive in the middle of the night on June 4, 1896 (Figure 1-1). During the same year, the Duryea brothers assembled the first thirteen automobiles for sale to the public in France. These early vehicles symbolize research, technological innovation, invention, creativity and pragmatism – some of the most important and fundamental tenets of today's automotive world.

Figure 1-1
Henry Ford's Quadricycle - 1896

In 1908, Henry Ford introduced the Model T and used the moving assembly line in 1913 to mass produce his vehicles. This increased the level of productivity allowed Mr. Ford to pay his workers $5.00/day, the highest wage of any manufacturing job at that time. Through these mass production techniques, automobiles were made affordable to a growing middle class. These new vehicles, powered by the internal combustion engine, allowed more people to settle away from the centers of cities, and goods to be transported to areas not served by trains or ships. The building of better roads became an imperative, leading eventually to turnpikes and national highway systems that ultimately changed the way in which our society was geographically distributed. It is interesting that the Gentleman's Cycling Club of New york promptly traded in their bicycles for Model-T's in 1913 (Figures 1-2 and 1-3)(Henry Ford Museum archives). Model T's became increasingly visible on city streets as illustrated by the photograph in Figure 1-3.

Figure 1-2
Gentleman's Cycling Club
(New York - 1913)

Figure 1-3
New Model T''s (New York - 1913)

Five years before the introduction of the Model T, another new technology-based industry was born. In 1903, Orville and Wilbur Wright successfully tested the first airplane at Kitty Hawk, North Carolina, powered also by an internal combustion engine. Henry Ford's first venture into aviation came in 1909 when he financed the building of a plane with a 28 hp Model T engine (O'Callaghan, 1993).

In 1926, the first flight of a liquid-fueled rocket occurred, followed in eleven years by the first working turbine jet engine. The number of airplanes and rockets has not approached the number of automobiles or trucks nor have they impacted individuals lives and the planet's ecosystems as much as the automobile. However, they allowed us to see the planet with more of a global perspective than what we could see from the automobile. These technologies have lead to the development of satellites, which have made possible a real-time perspective of our changing global ecosystem from space. As described later in this paper, satellites equipped with advanced sensors, imaging and communications systems will become an important technology to help solve the environmental and energy challenges of the 21st century.

The 20th century also saw the progression of the second phase of the industrial revolution, called by some the scientific or knowledge-based revolution. Through the accumulation of scientific knowledge, people have learned to harness matter and energy to develop entirely new materials; communicate with each other across large spaces; control machines and other inanimate objects; make mathematical calculations of immense proportions; handle large quantities of data efficiently; and even to transform matter into energy for useful purposes. In this century, we have seen new science lead to an explosion of new industrial and consumer products including the digital computer; the development of microwave and radio communications; and a whole new energy industry based on fission of atomic nuclei. From this "R&D" has emerged a better understanding of how molecular science is controlled; how living systems work and replicate themselves; how diseases can be controlled; and how the building blocks of matter can be assembled into a seemingly endless number of new materials that can be hard or plastic, sticky or crystalline, or can replicate images through processes that are initiated by light and other energy forms.

Much of this new technology has been utilized in the evolution of the modern automobile, which is now a composite of metal alloys, synthetic polymers, and computer controlled electrical and mechanical devices. It is powered by an increasingly sophisticated internal combustion engine whose fuel and air mixture is also computer controlled, and whose chemistry is understood with much greater certainty than could be imagined at the turn of the century.

These emerging technologies of the 20th century will provide the basis for the development of new technologies to help solve the many challenges that will face the automotive industry during the 21st century.

Approximately 47 million vehicles were manufactured in 1997 and the number of operating vehicles on the planet totals nearly 600 million (Graedel, 1998). The same knowledge base that created this incredible fleet has also revealed that its impact on the environment of our planet is significant. At the time new science was making the automobile more attractive and more affordable, it was also giving environmental scientists the tools to see that the automobile, and many of the other products of the industrial and scientific revolutions were not as benign as we had previously thought.

Many of today's possible solutions to the environmental, energy, and transportation challenges are founded on advanced concepts and ideas that were developed by transportation futurists and inventors in the early 1900's. Some of these early inventions included electric vehicles (Figure 1-4), hybrid electric vehicles (Figure 1-5), flying automobiles or "autoplanes", and "autoboats."

Figure 1-4
The First Electric Model T (1911)

Figure 1-5
The First Hybrid-Electric Vehicle (1912)

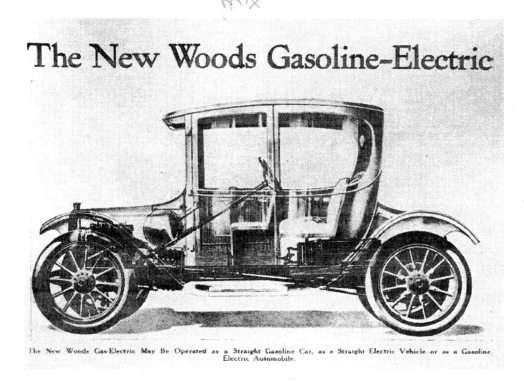

Several inventors developed hybrid versions of the automobile and airplane during the period between 1910 and 1920. Aviation pioneer Glen Curtiss built the first actual flying car in 1917 (Lee, 1983). Several autoplanes were developed during the 1930's in which the wing section could be easily detached. The Autoplane Company of America developed a three-wheeled autoplane during 1935-1939 with a rear-mounted propeller and long tail. Robert Schuetzle developed the Autoplane CR-2 at McClellan Air Force Base in Sacramento during 1940 for military operations (Figure 1-6). This 3-wheeled vehicle was designed so that the twin-engine wing section could be easily removed and re-attached. The wing section used twin Lycoming engines for air flight and a separate 6-cylinder Ford engine powered the car. Many other "autoplane" concepts have been developed since that time but they never became commercially practical because such vehicles were too heavy and they could not meet the increasingly stringent safety standards developed for the aircraft and automotive industries during the 1950's and later (Lee, 1983).

Figure 1-6
An Early Hybrid of the Automobile
and Airplane (1939-1940)

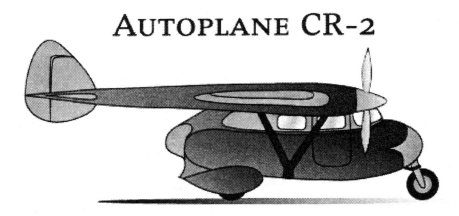

AUTOPLANE CR-2

POWERED BY TWO 75 HP LYCOMING 60-145-C ENGINES

"Automated highway corridors" were first conceived in the 1930's and working models were displayed at the 1939 World's Fair. The concept was that vehicles of the future would be able to travel to their destination in these corridors without driver intervention. Although this technology inspired the world and future generations of science fiction writers, there have been no compelling reasons to further develop such technologies, until recently.

Shortly after the introduction of the Model T, Ford chose a relatively new material, celluloid plastic, for the side curtains, which heralded the start of the use of lightweight plastics and polymer-based fibers in today's automobiles. During the 1920's work began on the development of synthetic fuels (or syn-fuels), especially fuels from coal (National Research Council, 1990). Henry Ford spearheaded research programs to develop renewable fuels and vehicle materials from soybeans (Bryan, 1990). Although many new technologies have been developed for the production of syn-fuels, they have not been cost competitive with petroleum-based fuels.

Many of these early inventions became "book shelved technologies" because of cost, complexity, difficulty of production, safety and other factors. Furthermore, the current drivers of fuel efficiency and low emissions were nonexistent. Even though prototype technologies such as fuel cell/electric powered vehicles do not appear practical as we approach the end of this centennial, such technologies may become commonplace within the next hundred years. Such innovative technologies will continue to shrink the world in time and space during the 21st century, but at a greatly accelerated rate.

For more than fifty years, technological progress in the automobile industry has been marked by a convergence toward a common set of basic technologies. As Henry Ford II said about thirty years ago, "When you think of the enormous progress of science over the last two generations, it's astonishing to realize that there is very little about the basic principles of today's automobile that would seem strange and unfamiliar to the pioneers of our industry." This is pretty much true today.

The effect of technological change on the automobile has been incremental, seeking continual refinement of the existing technology in order to increase productivity and reduce costs. Much of the innovation has been in the process of manufacturing rather than in the product itself. Competition in the automotive industry of the 1950's and 1960's manifested itself in less fundamental product alterations, such as styling and accessories.

The rate of technological innovation began to change rapidly during the 1980's and has accelerated through the 1990's. Such change has been referred to as "Epochal Innovation" and it is discussed further in Chapter 5, "The Global Revolution and The Automobile."

Chapter 2
Evolution of Environmental Awareness and Protection in The 20th Century

One cannot say that the industrial revolution and environmental awareness went hand in hand. Indeed, it was clear all along that the industrial revolution had a dark side. As cities grew to hold larger concentrations of workers for the factories, living conditions were often overcrowded and unhealthy, hygiene was poor, pollution increased, working conditions were hazardous, and some factories employed young children. Factories often omitted noxious smokes and odors from their stacks, discharged colored and toxic waters to their receiving streams and buried wastes that were clearly hazardous. All of this was tolerated by a populace that was glad to have jobs and by politicians who were indebted to both workers and management for their positions. Environmentalism, to the extent that it existed at all, was limited to a few conservationists and idealists who knew that there must be a better way.

As for the automobile, the initial concern for its effect on the environment was primarily local: a bit too much noise and some noxious smoke, but nothing serious. As urbanization and the popularity of the automobile increased, however, it became clear that internal combustion vehicles had more than a local effect on the environment. The severe smog episodes first observed in Los Angeles during the 1950's were found by Haagen-Smit to be due primarily to the photochemical reaction of automotive emissions in the atmosphere (Seinfeld, 1986). Over the next forty years, the chemistry of this complex process would be mastered, but it was far from clear how to avoid it, given the love that Los Angeles citizens have for their automobiles (Lents, 1993).

During the 1960's and 1970's, the tools of science were also beginning to be turned toward other emissions of our industrialized society. Fish kills could be traced to effluents from factories and untreated municipal sewage, sometimes far upstream from the site of the damage (Carson, 1962). Emissions of sulfur and nitrogen oxides from factories were found to cause damage to forests and lakes at locations quite far from the origins of these pollutants, sometimes even in a different country. Pesticides and other industrial chemicals were found to be accumulating to alarming levels in fish and other animals.

The environmental movement began to gain momentum in the 1970's and over the next two decades a series of laws and regulations were codified that brought government control over industry to a level which was unprecedented in modern times in western society (Wolf, 1988). Approximately 50 major environmental regulations have been promulgated in the United States in the past 100 years (Graedel, 1998) (Figure 2-1).

Figure 2-1
The Evolution of Command and Control Based Environmental
Regulations In the United States

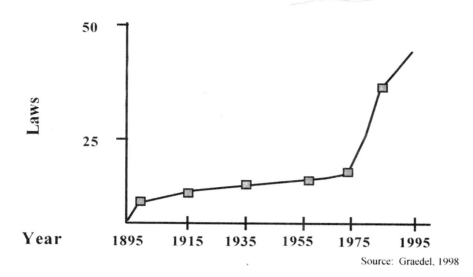

Source: Graedel, 1998

Many have claimed that there are now too many laws and regulations in the U.S. and that regulation is stifling the economy. While there is some truth in this argument, there is also good evidence that enforcement of selected regulations has been a necessary component of environmental protection. Good laws have had a beneficial effect on the environment and public health. Since the first passage of the Clean Air Act in 1970 (Wolf, 1988), vehicle emissions such as nitrogen oxides, hydrocarbons and carbon monoxide have been reduced by more than 96%. Among the success stories, the control of lead in gasoline has been the most impressive. A dramatic reduction in airborne lead levels in U.S. cities as a result of the implementation of unleaded fuel resulted in a rapid decline in the levels of lead in children's blood (Commission on Life Sciences, 1993) (Figure 2-2).

Figure 2-2
Lead Blood Levels as a Function of Ambient
Air Lead Concentrations in the United States

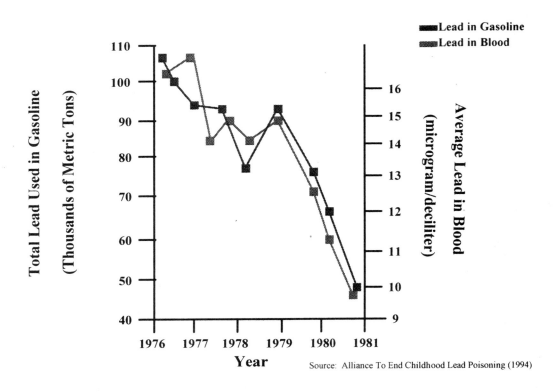

Source: Alliance To End Childhood Lead Poisoning (1994)

As the environmental movement was taking hold, research also began to reveal that the environmental impacts on the planet were not just local but regional and global. New scientific data, generated from the satellites and sophisticated sensors, that were themselves products of the industrial and scientific revolutions, showed that anthropogenic emissions were affecting the control systems that regulate our planet.

The theory that chlorofluorocarbons could react with ozone in the troposphere was initially discredited, but it has since been scientifically verified through atmospheric and laboratory studies (World Resources Institute, 1993). The ozone hole was one of the first clear pieces of evidence that the effect of human industry accrued on a global scale. Since the 1970's, an ozone hole has formed over Antarctica each Fall, in which approximately 60% of the total ozone is depleted (Figure 2-3) (NASA, 1998).

Figure 2-3

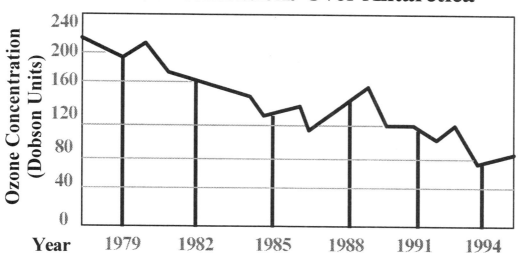

Ozone Concentrations Over Antarctica

Source: NASA

The world's population has grown at a rapid pace during the past 100 years to a current level of 5.98 billion. It is increasing by approximately 200,000 people per day or the equivalent of a new megacity the size of Bangkok every 2 months. The population is projected to increase to approximately 8.3 billion by 2025 and 10 billion by 2050 (Figure 2-4) (US Bureau of The Census, 1997). Most of the world's habitable areas will have a high population density by that time, which will increase the need for private and public transportation.

Figure 2-4

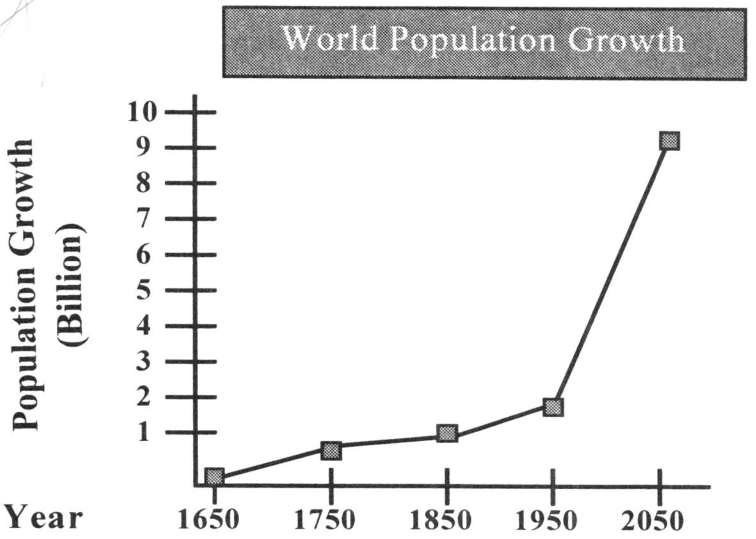

World Population Growth

Source: U. S. Bureau of the Census

The transportation industry and its products can have a significant effect on the depletion of the earth's resources. Vehicles consume approximately half of all the crude oil produced world-wide and that share has been increasing. The current global vehicle fleet of 600 million vehicles will emit 4.0 billion tons of carbon dioxide into the atmosphere during 1998 (World Resources Institute, 1993). This is approximately 20% of the total quantity produced by all human activities. The number of vehicles on the road is growing currently at a rate that is twice as fast as the world population growth and it is projected to rise to 1.4 billion vehicles by the year 2050 (World Resources Institute, 1996).

The growth in popularity of the automobile has been driven by economic prosperity and by the growth of megacities, as the Earth has become rapidly urbanized. About 75% of the inhabitants of the mature countries live in cities, compared to a global average of 50%. The fastest rate of urban growth is in the emerging countries, resulting in problems of air and water pollution, waste disposal, energy resources, and traffic management.

By 2050, about 15% of the earth's population is expected to live in 60 megacities above 10 million in population. All available evidence suggests that these people will be enamored with the automobile as much as those who now own vehicles (World Resources Institute, 1996) .

Governments will face pressure to build and maintain the infrastructure to support these vehicles. Roads will cover a significant proportion of the surface area in cities and they will increasingly consume forest and farm land since very few places in the world have established guidelines to limit the growth of roadways and development in rural areas. Satellite imaging of undeveloped areas of the earth vividly demonstrates the effect of roads and the automobile, even in remote regions. Figure 2-5 shows a satellite image of the Amazon jungle, near Santa Cruz, Bolivia.

Figure 2-5
Satellite Image of the Amazon jungle

Image Source: ERIM International (1998)

This image shows a "fish-bone or "feather" deforestation pattern typical throughout the Amazon. Vegetation along the road is systematically slashed to permit agriculture as access roads are carved into the jungle (World Resources Institute, 1993). As the soil is insufficiently fertile to support intensive agriculture for any length of time, paths are carved perpendicular to the original roads to provide access to more land and the pattern repeated until the fishbone road pattern emerges.

The rapid expansion of the global road networks will allow new population centers to form. Figure 2-6 illustrates the global population distribution for the years 1750, 1950 and 1998. Each point in this model represents one million people within a surface grid of 10,000 sq. miles. During the next century, the population will continue to disperse throughout the globe as transportation and communication systems improve.

Figure 2-6
Global Population Distribution
Actual for 1750 and 1950 and 1998
(Legend - Greater than One Million People/Point)

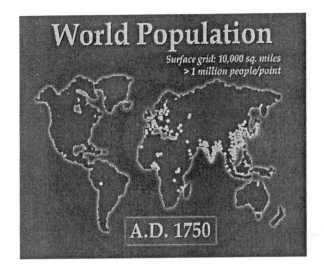

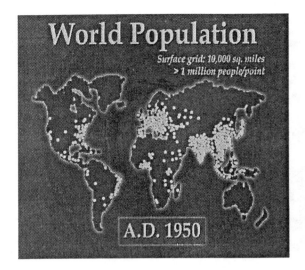

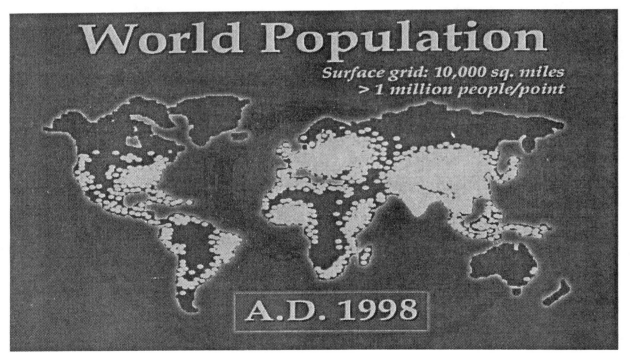

Reference: Internet Sites
http://www.wpfvf.com/indexframe.html
http://www.physics.iastate.edu/gcp/issues/pop/popmoviev7.flt.movpop/growth.html

During the past century, we have seen that efficient transportation contributes to economic growth, increases the standard of living and benefits the public. This is why many emerging countries, such as China, have designated the automotive industry as a "pillar industry" for the future - a key sector of commerce that is necessary for the social and economic development of a country. The automobile will continue to play a major role in improving the lives of people throughout the 21st century.

World industry, as measured by energy utilization and gross domestic product (GDP) per person, is growing rapidly. As a result, we will continue to be faced with a delicate balancing act. On one hand, a careful balance will be needed to achieve economic success for the global population, while on the other hand, the environmental quality of the planet will need to be preserved.

Chapter 3
Environmental and Energy Challenges
of the 21st Century

As we enter the 21st Century, most of the world will continue to grow at a rapid economic pace, a trend that will be spurred on by global competition. Several factors have contributed to this boom including the opening of new markets in countries that were formerly closed, the flow of capital into emerging countries, and the globalization of financial institutions, manufacturing and service industries. Emerging markets in South America, Asia, Africa and Oceania are now competing with the industrialized nations, producing and consuming vast new quantities of commodities. In Shanghai, Bangkok, Kuala Lampur or virtually any East Asian commercial center outside of Japan, large scale construction projects, elaborate shopping malls, five star hotels and traffic jams are becoming commonplace. Equally important, individual citizens from around the world are demanding more and more consumer products, including automobiles, that are products of the industrial and scientific revolutions.

The result of this boom is becoming as apparent in the emerging markets as it has been in the mature markets during the past century. For example, during the 20th century many Western cities developed an urban and suburban configuration where mobility depended primarily on private automobile use. However, this strategy led to increasing levels of traffic congestion, air pollution, resource depletion, and urban sprawl. Asian mega-cities such as Bangkok and Beijing have been growing in the same mode. It has become apparent that pollution and environmental problems will increase in these regions until their economies develop further and long-term plans are established for environmental protection.

Although, the residents of the emerging nations place a high priority on economic development and good jobs, they are not as willing to disregard the environment as their counterparts in the western nations did during the first half of the 20th century. These attitudes are exemplified by numerous studies such as a recent survey in Beijing, which shows that eighty-nine percent of the residents believe that the city has serious environmental problems. They enumerate air pollution, solid waste, water contamination and noise as the top four environmental problems that should be immediately addressed (China Daily, Oct. 27, 97).

We have identified three major environmental challenges to the automotive industry during the next 100 years as outlined in Table 3-1.

Table 3-1: Some Key Environmental Challenges of the 21st Century

Last 100 Years	Next 100 Years - Challenges
Regional and National Environmental Focus	**Global Environmental Focus**
Resource Utilization	**Resource Use Efficiency**
Environmentally Reactive	**Environmental Stewardship**

Successful automotive companies of the future will need to develop 1) a global environmental focus, 2) products and processes that help achieve an efficient use of resources and 3) policies, methodologies and technologies that support global, environmental stewardship.

Global Environmental Focus

Until about 30 years ago, the concern for the environmental effects of vehicles on the environment was primary local. The use of advanced analytical measurement technologies, such as environmental satellites, have allowed scientists to track the dispersion of air pollutants over hundreds of miles across the surface of our planet (NASA, 1998). As a result, it is generally accepted that environmental pollution can have regional as well as global ramifications. The production and use of vehicles results in the global distribution of greenhouse gases, nitrogen oxides, hydrocarbons and other pollutants. However, the automotive industry is by no means the primary source of these emissions.

Recent satellite images vividly demonstrate the potential global effects of air and water pollution from a multitude of sources. The recent forest fires in Southeast Asia, Mexico and Florida have affected thousands of square miles and almost a billion people in these regions. A satellite photo (Figure 3-1) of the fires in Indonesia provide a graphic example of how airborne particulate matter can affect a large region. The amount of greenhouse gases emitted by these fires will be greater than that emitted from all fossil fuels consumed in Europe during 1998. Other satellite images demonstrate that emissions from the many coal-fired power plants in Shanxi Province, China can cover a region that extends to Japan.

Figure 3-1
Satellite Photo of Smoke from the
1998 Forest Fires In Indonesia

Resource Use Efficiency

One of the major challenges of the 21st century will be to improve the efficient use of the earth's resources. Sustainable development is a common terminology that has been used to help describe resource use efficiency. A generic definition of sustainable development is to protect the environment, promote economic growth, and provide economic opportunities for current and future generations. As far as sustainable development is concerned, all countries are emerging countries. Sustainable development will require that all nations help to ensure that the ecological, economic and social bases and principles of life are guaranteed and improved in ways that are both durable and sustainable in the future (Hart, 1996).

Ford is committed to the efficient use of resources and it has developed strategies to support such policies. Ford uses the terminology, "environmental stewardship" to generally describe these policies. This is a proactive approach to preserve the global environment, promote the development of high fuel efficient vehicles, use materials efficiently, and promote environmental awarness through education. It is important that all automotive companies ultimately develop a strategy of "environmental stewardship" in which common standards and guidelines are adopted globally. As outlined later, it is not expected that this global environmental harmonization process will be fully implemented until about 2020.

The governments of most emerging countries are keenly sensitive to environmental concerns. Dr. Song Jian, former Chairman of China's State Science and Technology Commission and a member of the State Council, has been one of the key leaders in China's policies of sustainable development. He developed mathematical relationships between China's rapid growing population and China's resources during the 1970's. The popularization of his theories was facilitated by the publication of his popular book on this subject (Song, 1985) and the implementation of China's "Family Planning Policy" in 1979.

This policy has been widely accepted by most people in China. As a result, it is becoming possible for Chinese families to focus on education, the development of business opportunities, the attainment of better living conditions, and incidentally the purchase of vehicles in increasing numbers. Although China's population currently represents about 1.25 billion (21%) of the world's total, that proportion is expected to drop to about 19% or 1.75 billion people by 2050 as a result of such policies.

History has shown that as the economic viability of families improve, the size of the family will decrease. As the global trend of personal wealth increases, the resulting smaller family units will ultimately affect the type of personal use vehicles that will be needed in the future. This factor combined with the high population density of the emerging megacities of the future, implies that these families could desire smaller vehicles.

Global Energy Resources

The mature nations now account for 70% of the world's energy use (Figure 3-2). Emerging nations will need to increase fossil fuel imports to meet demand as their economies develop. The emerging markets of Asia already consume some 25% of the world's annual demand for basic commodities and these economies are now growing at 2-4 times the rate of the U.S. Devleoping countries could raise real gross domestic product per person by 270% by 2020 if their market reforms continue at this rate. China could have the world's largest economy by 2020, representing about half of the combined annual outputs of all the economies in the mature markets.

As a result, China will become the largest consumer of energy in about 2030. It is projected that the mature and emerging nations will reach the same level of energy consumption (Figure 3-3) by 2035. Since more than 80% of the current global energy use is provided by fossil fuels, Figures 3-2 and 3-3 also provide a rough approximation for global contributions of carbon dioxide emissions.

Figure 3-2
Global Energy Use
Year 1998

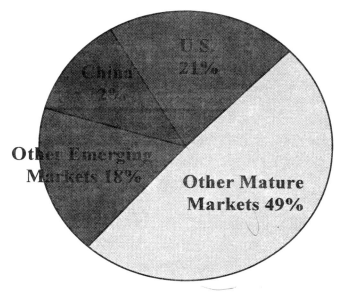

Figure 3-3
Global Energy Use
Year 2035

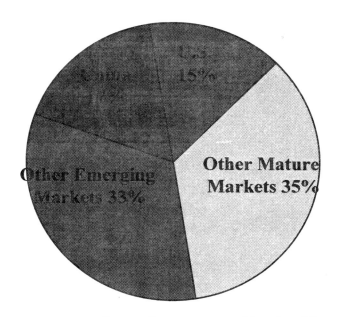

Source: Intergovernmental Panel on Climate Change (1997)

During the past twenty years, the world's major energy companies have carried out extensive exploration for fossil fuels on a global scale. As a result, world energy reserves have increased significantly, despite rising levels of consumption (World Resources Institute, 1996). Petroleum reserves are of particular importance, especially to the transportation industry. Petroleum supplies 40% of the world's total energy requirements, accounting for about 22 billion barrels per year.

Studies of energy resources incorporate both geological and economic factors, and both approaches can yield different perspectives. At present, there is little near-term concern over petroleum supplies, since production is ample and oil prices are relatively low. The world's fossil fuel resources are finite, however, and global production will eventually peak and then start to decline.

Forecasts of when petroleum production will decline have been made since the 1940's. The estimate at that time was 600 billion barrels compare to 2,300 billion barrels today (Campbell, 1995). It is widely acknowledged by petroleum geologists that current estimates are reasonably accurate due to the availability of sophisticated exploration technologies and extensive global exploration that has taken place during the past two decades. Therefore, more accurate predictions can be made for when most of the earth's fossil fuel resources will likely be expended. These predictions are presented for the World and China to help illustrate the dynamics of this subject.

Table 3-2 summarizes the known global resources of coal, petroleum and natural gas. The readily recoverable reserves are those that can be extracted under present and expected local economic conditions with existing available technology. The remainder of the fossil fuel reserves cannot now be recovered under present economic conditions and/or the technology does not exist to extract them. However, past experience suggests that new technologies will become available to allow the viable recovery for some of these resources.

Table 3-2:   A Comparison of Global Fossil Fuel Reserves and
             Readily Recoverable Fossil Fuel Reserves

	Estimated Fossil Fuel Reserves	
	Total	Readily Recoverable
PETROLEUM (Billion Barrels)	2,950	2,300
NATURAL GAS (Billion Cubic Meters)	212,000	141,400
COAL (Billion Tons)	1830	1030

Source: World Energy Council and the World Bank (1993), IEA (1998).

The estimated lifetimes in years for the recoverable fossil fuel resources are determined by dividing the total known resources by the current rate of use (consumption per year).   These lifetime estimates will remain reasonably accurate as long as the increased requirement in energy resource used each year matches the increased increment in new reserves identified.

$$\text{Lifetime of Reserves (years)} \quad = \quad \frac{\text{Known Resources}}{\text{Consumption/Year}}$$

<u>Assumption</u>: Increased Demand/Year = New Reserves Identified/Year

This hypothesis currently provides an optimistic lifetime for these reserves since the current world demand for coal, petroleum and natural gas has been growing annually since 1989 at a percentage rate that is higher than the percentage rate of growth in reserves.

Although it is expected that the U.S. and other mature markets will decrease energy utilization per unit GDP, the rapid growth of China and other emerging economies will lead to a more rapid depletion of the world's energy supplies, resulting in global projected lifetimes of approximately 39 years for petroleum, 60 years for natural gas and 221 years for coal (Figure 3-4). These estimated lifetimes would change for the three major fossil fuels as the demand for each shift throughout the next century. For instance, the rate of consumption of natural gas will increase in part because it is an environmentally friendly fuel.

Figure 3-4
Estimated Lifetime (Yrs.) of Recoverable Fossil
Fuel Reserves for the World and China

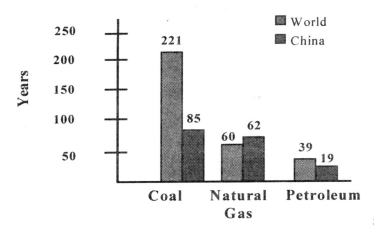

Source: U.S. Department of Energy

Recent studies estimate that global petroleum production will peak somewhere between 2010 and 2025 (World Energy Council, 1995; U.S. Dept. of Energy, 1995; Masters, 1991). Even after the peak, oil production will continue, although at a declining rate. Other fuels would fill the gap, for example, natural gas, coal, heavy oils and bitumen can be used to produce liquid or gaseous substitutes for crude oil (Masters, 1991).

China will deplete its presently known domestic reserves of petroleum in 19 years and natural gas in 62 years, respectively, unless domestic supplies are conserved and petroleum imports are increased (Table 3-3 and Figure 3-4). In order to make up for this shortfall, China's oil imports will need to quintuple by the year 2010 to over 3 million barrels per day. That level will represent one-half of Saudi Arabia's current production and 15% of the total global consumption. Already China is moving very quickly to buy reserves of oil from the Middle East and natural gas reserves from Russia (Asia Society, 1998; McElroy, 1998).

Table 3-3: China's Energy Reserves[1] and the Estimated Lifetime of These Reserves[2]

	Total	% of World Reserves	Consumption Level	Estimated Lifetime of Reserves (Yrs.)
PETROLEUM	24 (Billion Barrels)	2.4%	1.26 (Billion Barrels/year)	19
NATURAL GAS	1161 (Billion Cubic Meters)	0.83%	18.8 (Billion Cubic Meters/year)	62
COAL	126 (Billion Tons)	11%	1.48 (Billion Tons/year)	85

[1] U.S. Dept. of Energy (1996)
[2] Assumes that the increased rate of energy resource use will match the increased rate of discovery

A doubling of the proportion of people who reside in Chinese and Indian cities would increase per capita energy consumption in those countries by 45%, even if their industrialization and income per capita remain unchanged (U.S. Dept. of Energy, 1995).

As the global supplies of petroleum and natural gas dwindle, coal-rich countries such as the U.S., Russia, Canada and China will rely more on coal as their primary feedstocks for fuels, energy and carbonaceous chemicals. During the later part of the 21st century, the coal-rich countries will become the key energy and carbonaceous materials exporters to the rest of the world. It is important to remember that coal is not an environmentally benign energy source, and that the development of clean coal technologies will be needed. Unless the growth in global demand for energy is significantly reduced and the mix of energy source utilization drastically changes, nearly all fossil fuel reserves will be depleted by the end of the 22nd century.

The ultimate depletion of the fossil hydrocarbon reserves will require that environmentally acceptable alternative energy technologies be more widely developed. One challenge will be to develop nuclear energy as an environmentally acceptable energy source (Kiefer, 1979). France and Japan have had the vision to proceed with the development of safe nuclear power generation. Nuclear power generation accounts for about one-fifth of the electricity generated in the U.S., whereas France derives about three quarters of its electrical power from this source. Currently, all nuclear power plants generate energy by nuclear fission, however during the next century, breeder reactors could become more prevalent followed by nuclear fusion. It should be noted that there remain many technical and political issues associated with nuclear power production, including waste disposal and potential terrorist activities.

Whatever the global energy scenario, it is clear that the global fleet of approximately 1.4 billion vehicles by about 2030 will need to begin utilizing alternative sources of fuels and energy.

Electricity is one primary alternative source of energy with great promise for the automotive industry. Although there are widespread technical and affordability issues, its widespread availability will probably lead to the prevalence of electric powered vehicles, assuming these vehicles become performance and cost competitive with gasoline and diesel fueled vehicles.

The average global efficiency for the conversion of fossil fuels to electricity is about 32%. The efficiency of electrical production averages about 36% in the U.S. compared to about 27% in China. Electrical generation using renewable energy sources represents about 19% of the world's total, 95% of which is hydropower. Future technical developments in the area of renewable energy sources appear to hinge on the development of more efficient solar photovoltaic cells and the more efficient production of biofuels. Lacking these alternatives, fossil fuels will continue to be attractive throughout the 21st century in spite of their potential environmental impacts.

It is possible now to economically convert natural gas to electricity with an efficiency of over 50%. Some of the waste heat generated in the process can be inexpensively converted to useful steam, raising the efficiency to nearly 60%. It is possible that the conversion efficiency of fossil fuels to electricity could approach 50% globally by 2100. This drive toward a high conversion efficiency may be facilitated by the establishment of energy agreements between the world's nations to work together and share power generation technologies to achieve these goals.

Environmental Stewardship

Farmers who work the land appreciate the concept of the efficient use of resources. Henry Ford was born and raised on a farm and he practiced and understood the need to sustain the land. As a businessman, Henry Ford was an early champion of recycling.

Although this is a popular cause today, it was hardly a trendy management concept 60 years ago.

Recycling is now a popular ethic. Today, the automotive industry is making considerable progress in curtailing the environmental impact of its activities such as reducing and avoiding pollutants at their source, by recycling, by energy optimization of manufacturing facilities and by replacing environmentally harmful products such as cholorofluorocarbons.   However, because of the high potential environmental impact of its processes and products, and the ongoing trend toward globalization, the automotive industry must assume a major responsibility in motivating resource use efficiency (Graedel, 1998).

Recycling in the automotive industry is becoming an important constituent of resource use efficiency. In the mature markets, more than 80% of the materials from disposed vehicles are being used to produce new vehicles and other commercial products. The future trend will be to incorporate a higher percentages of recycled materials from other sources to produce vehicle components. Figure 3-5 shows some of the recycled materials that Ford is already using to produce their vehicles.   As this approach is adopted in the future, the global automobile industry will be able to approach a zero net utilization of materials.

Figure 3-5

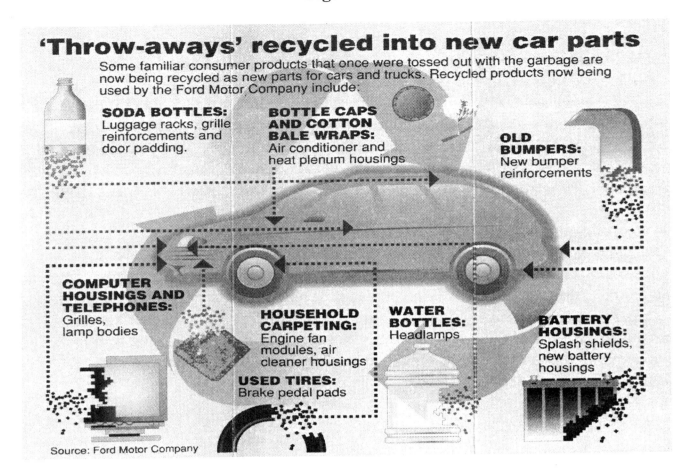

'Throw-aways' recycled into new car parts

Some familiar consumer products that once were tossed out with the garbage are now being recycled as new parts for cars and trucks. Recycled products now being used by the Ford Motor Company include:

SODA BOTTLES: Luggage racks, grille reinforcements and door padding.

BOTTLE CAPS AND COTTON BALE WRAPS: Air conditioner and heat plenum housings

OLD BUMPERS: New bumper reinforcements

COMPUTER HOUSINGS AND TELEPHONES: Grilles, lamp bodies

HOUSEHOLD CARPETING: Engine fan modules, air cleaner housings

USED TIRES: Brake pedal pads

WATER BOTTLES: Headlamps

BATTERY HOUSINGS: Splash shields, new battery housings

Source: Ford Motor Company

Ford Chairman and CEO Alex Trotman, stated at a National Press Club event in 1997 that "being an environmentally friendly company is a business imperative. It is an integral part of meeting the needs of our customers. Certainly the people at Ford consider themselves environmentalists since every one of us is concerned about the health of our families and the condition in which we leave the planet for those who follow."

Most major international companies have adopted a corporate policy of being environmentally proactive. This is in part because more people in the mature markets are demanding that products and manufacturers are environmentally friendly. The residents of the emerging nations are not far behind.

Chapter 4
Future Environmental
Approaches and Methodologies

The automotive industry and other major industries and businesses, are entering a new wave of globalization with the arrival of the "Mega Competition Era". Automotive companies now realize that global management and technology strategies are crucial for survival in the world's automotive industry. Global strategies will need to be developed and implemented for product development, manufacturing, marketing and environmental stewardship.

Successful automotive companies of the future will need to develop and support many new environmental approaches and methodologies (Table 4-1). Some key imperatives of the 21st century will be to 1) advocate global approaches to the establishment of international standards, 2) champion consortia in cooperation with government, academia and industry, 3) develop new approaches that integrate technical, social and economic solutions, while balancing public and private approaches, 4) support research and development as needed to help develop sound environmental policies and 5) utilize advanced approaches and methodologies, such as life-cycle assessments to help evaluate the potential environmental, economic and energy consequences of the automotive industry on a global basis.

Table 4-1: Future Global Approaches to Environmental Stewardship

Last 100 Years	Next 100 Years - Challenges
Regional Approaches	**Global Approaches**
Single Sector Approaches	**Industry, Government and Academic Consortia**
Technical Solutions	**Integration of Technical, Social and Economic Solutions**
Political Basis for Environmental Policy	**Scientific Basis for Environmental Policy**
Single Basis Assessment	**Life-Cycle Assessment**

Multi-National Approaches

As industrial globalization unfolds, automotive companies in the emerging markets will become increasingly competitive with those in the mature markets. Therefore, it is to the collective advantage of all major global automotive companies to work with national governments to achieve a certain level of global harmonization of standards while still accommodating national preferences.

One factor that will assist in environmental protection is the trend toward global harmonization of industry standards. Global harmonization of automotive engineering specifications and regulatory standards are key focus areas for automotive companies and their suppliers. This is because there are a wide array of automotive specifications and standards, worldwide.

Some countries develop specific engineering specifications and regulatory standards to create "hidden" trade barriers to the flow of goods, capital and technology across international borders. These disparate national environmental specifications and regulations result in a need for manufacturers to develop different variations of a vehicle with a concomitant waste of energy and materials. It has been estimated that vehicle costs could be reduced by as much as 12% if all nations worked together to develop and implement the global harmonization of automotive specifications and standards.

The Montreal Protocol, negotiated in 1987 to ban ozone-depleting freons, is an excellent example of a successful international environmental agreement. It provided market mechanisms for lowering costs, offered compensation for participation by emerging countries and achieved nearly universal global participation.

During the 1990's several international organizations formed working groups that continue to play an important role in the establishment of global environmental policies and technologies for the automotive industry. For example, the International Standards Organization and the World Health Organization have been instrumental in their efforts to harmonize global standards for the protection of public health from environmental hazards.

Since the Rio de Janeiro Earth Summit of 1992, the world's governments have been building a foundation of international environmental laws that will help meet the environmental challenges of the 21st century. The Summit experts established principles that protect the environment while allowing the development of the global economy. After the Summit, several international organizations formed working groups which will continue to play an increasingly important role in the establishment of global environmental policies and technologies for the automotive industry.

The World Trade Organization (WTO) has focused its efforts on tariffs to reduce various types of trade barriers, especially for the automotive industry, and the World Health Organization (WHO) has established limits for human exposure to toxic and hazardous chemicals.

Likewise, the voice of the manufacturer is being expressed through global automotive organizations such as FISITA and SAE International. Both organizations continue to take leadership roles in helping the automotive industry establish a policy of global environmental stewardship.

The environmental management system developed by the International Standards Organization, ISO 14001, is one of the most successful examples of multi-national approaches to the establishment of environmental standards. This system covers environmental performance evaluations, audits and product life-cycle assessments (ISO, 1997). It is not legally binding and does not include emission limits or standards, but these voluntary, primarily procedural standards aim to improve corporate environmental performance by establishing a single, uniform set of internationally accepted measures against which firms can gauge their environmental procedures (Lamprecht, 1997).

Many European countries now require that automotive companies must become ISO 14001 certified to sell vehicles. China is committed to its implementation, countrywide. This standard has been instrumental in helping Ford improve its efforts to protect air, water, land and natural resources in its manufacturing facilities, worldwide.

There are a number of hurdles to overcome before a global approach to environmental protection and resource use efficiency can be fully achieved. Some believe that emerging market nations should be allowed to reach a certain level of gross domestic product and international competitiveness with the mature nations before they begin to seriously address the issue of national and global environmental issues. Others feel that this is a formula for disaster and that mature nations should boost aid and provide more environmentally friendly technology to emerging countries (Bender, 1991).

There is substantial debate within the trade and environmental communities about how to cope with uneven environmental regulations that may affect the flow of goods, capital and technology across international borders. One of the problems with international environmental agreements is that each country would prefer that all others reduce their emissions, but each has little incentive to reduce its own emissions unilaterally. Any workable agreement will need to insure that all countries will collectively benefit economically, and that environmental impacts will not be unfairly distributed. Such agreements should make the aggregate gains from cooperation as large as possible. This could require some kind of market mechanism, either a tax on emissions or some kind of program for trading in negotiated emission entitlements.

It has been found that almost all countries comply with globally established treaties if those treaties have provisions that provide for the restraint of trade with the non-compliant countries. The biggest problem that will face the establishment of international environmental treaties is sometimes called "free-riding." If the mature markets reduce their emissions, the cost of their products will increase. Competitive advantage for the pollution intensive industries could shift to the non-participating industries. These countries will thus increase their output and increase their emissions as a direct consequence of the abatement undertaken by participating countries.

Future efforts will be needed to determine under what conditions international standard setting is likely to be the most effective with respect to environmental protection and the maintenance of competitiveness.

Industry, Government and Academic Consortia

The environmental issues that the automobile industry will be facing during the next 100 years will be extremely complex. It is important that sound science be used to develop effective environmental policies. Scientists and engineers from industry, government and academia will need to work more closely with policy makers in the future to help accomplish this goal.

Workable solutions to the complex problem of reducing the environmental impact of the automobile will require the efforts and expertise of professionals from federal, state and local regulatory agencies, environmental groups, industry, academia and research organizations.

Ford Motor Co., General Motors Corp. and Chrysler Corp. established the Progress for a New Generation of Vehicles (PNGV) consortium in 1993 to develop advanced automobiles that will be up to three times more fuel-efficient than today's vehicles but safe, environmentally friendly and affordable. In addition to the automotive manufacturers, this consortium includes about 350 other American technology based companies and 19 Federal government laboratories. The objective is to develop a car than can travel up to 100 kilometers on 3 liters of gasoline (up to 80 miles/gallon) without affecting cost, performance, safety or emissions. It is expected that cooperative programs such as the PNGV will be globally expanded to other energy-intensive industries throughout the 21st century.

A number of cooperative efforts are being established between mature and emerging nations to address the future challenges of energy and the environment. An example is the U.S. - China Science and Technology Agreement that was signed in Oct., 1997 between the U.S. Dept. of Energy (DOE) and China's State Science and Technology Commission. This agreement is part of a global strategy by the DOE to increase R/D efforts in advanced fossil fuels, renewable fuels, energy efficiency, and fission and fusion nuclear technologies.

Cooperation between the automotive industry and academic institutions have also become more commonplace recently, especially in areas such as air pollution modeling, design for the environment, and related technical areas. In the future the automotive industry must encourage research universities to turn their full multidisciplinary capabilities toward the complex issues discussed above, in cooperation with industry.

Integrating Economic, Technical and Social Solutions

Governments and industries face complex choices and tradeoffs in choosing technical approaches to resource use efficiency in order to meet mounting social demands without compromising ecological balance and the health of present and future generations. New approaches will be needed that go beyond the command and control environmental policies of the past (Ginsburg, 1980).

Since these issues are very complex, advanced computer-based models will be needed that provide government policy makers with a multi-discipline systems approach to urban planning and the optimization of transportation systems in high-growth urban areas. Multi-discipline teams of experts who cover the areas of environmental sciences and engineering, economics, urban planning, automotive engineering, government policy and social sciences will be needed to develop such models. Input from anthropologists, humanists, and even philosophers will also help guide discussions toward more acceptable solutions. Such approaches will help guarantee a people-friendly urban environment that provides an efficient urban and inter-urban transportation system while achieving a goal of optimizing energy use and reducing the emission of pollutants.

Most economists agree that market-based measures such as a energy taxes or emissions trading programs could be more efficient and less expensive than command and control regulations which require emissions be cut to a certain level. One mechanism under consideration is the promulgation of an energy tax to encourage fuel and energy conservation, to help develop and implement alternative fuel and cleaner energy sources, as well as improve the infrastructure for public transportation. The residents of core urban areas would benefit by the increased availability of public transportation. This tax could result in an increased demand for smaller, more fuel-efficient vehicles, which will lead to higher profits. This tax could have positive economic, technical and social benefits if efficiently managed (Table 4-2).

Table 4-2: The Potential Benefits of an Energy Tax to Society and the Advancement of Transportation Technologies

Economic	Technical	Social
Lower Transportation Cost	Public Transportation	Improved Mobility
Improved Commercial Efficiency	Clean and Renewable Fuels	Improved Urban Environment
	Alternative Fuel Infrastructure	

International pollution trading programs could be increasingly used in the future to help reduce global emissions. These trading programs would reward companies and nations that develop innovative pollution abatement methods. This approach could be expanded from the successful trading programs developed for American utility companies to reduce the emissions of acid rain forming pollutants.

An international emissions trading scheme could be developed so that companies or nations that emit less of a particular pollutant than required could sell their emissions credits to those that are not able to meet their emissions limits. In this manner, a global limit on emissions (or energy use) would be ensured. For example, a country could try to reduce its carbon dioxide emissions to a prescribed limit, buy the right to emit more carbon dioxide, pay to preserve forests, or transfer technology to improve energy efficiency.

Trading schemes of this type will not be sufficient. However, it is clear that many countries will need to develop incentives and subsidies for energy efficiency that are both mandated and market driven and more energy efficient vehicles will be needed that are cost competitive on a "life-cycle economics basis."

The standards that drive these trading mechanisms should be fair to the developed as well as emerging countries, based on factors such as energy consumed per unit of Gross Domestic Product (G.D.P.). As an example, Figure 4-1 provides a trend of energy consumed (in British Thermal Units - B.T.U.'s) per dollar of G.D.P. for the United States. The average value in 1962 was 19.0 compared to an average value of 13.6 from 1990 to 1996. Although the U.S. has become more energy efficient since 1962 (at which time the value was 19.0) and the country has become slightly more energy efficient throughout the 1990's, much greater improvements will be needed throughout the 21st century if the U.S. is to reach the standards needed for a sustainable society.

Figure 4-1
Energy Consumed (B.T.U.'s) per Dollar of
GDP for the United States (1992)

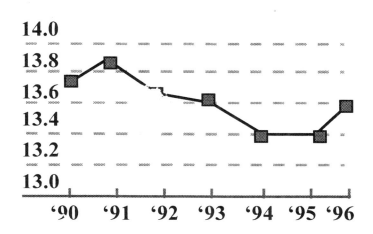

Source: Energy Information Admin., 1997

Scientific Basis for Environmental Policy

The improvement in the ambient air and water quality of the mature nations has been achieved over the past three decades by the promulgation of some well-defined and scientifically credible environmental regulations by the U.S. EPA, the World Health Organization and other regulatory agencies during the 1970's and 1980's. However, the environmental movement has always been driven by public pressure to act, even in the face of scientific uncertainties. Lately, the reasoned doctrine of using sound science to help set global environmental regulations appears to be challenged openly by those who feel that action cannot wait for scientific certainty. Unfortunately, this tension has been exacerbated by some members of the scientific community and the popular press who have become too anxious to draw conclusions that are based upon preliminary and/or questionable data. Too much emphasis is placed on singular cause and effect analyses, even when it is recognized that all natural processes result from complex, multi-causal origins. As poet Emily

Dickinson wrote more than 100 years ago, "belief without evidence is dangerous." That same advice applies today to the publication of environmental reports and the development of environmental policies. Several recent case studies are summarized to illustrate this point.

Benefit-Cost Assessment - The benefits of environmental programs are difficult to quantify and as a result cost-benefit assessments are not often included in justification criteria for environmental policies. Government spending on the environment by the mature markets rose rapidly during the past 25 years and such funding is expected to level off in the future. The emerging markets have a different problem. They need to continually increase funds to help implement environmental protection programs, while maintaining a certain level of growth to sustain their developing economies.

The future trend will continue to be toward pollution prevention and energy conservation. Energy conservation makes good economic sense given the costs of new power plants and the finite supply of fossil fuels. The automotive industry has been actively seeking ways to reduce the generation of waste. Such efforts have resulted in significant materials costs savings and reduced disposal costs (Vig, 1994). As this process continues, it is inevitable that future benefits will eventually decline relative to costs for pollution prevention or energy conservation. Decisions will become more contentious as the cost/benefit ratio increases, which clearly emphasizes the importance of greater scientific certainty in the risk assessment process.

Some form of risk-based priority setting will be essential if environmental regulation is to make economic as well as environmental sense. The use of more resources, than are absolutely necessary to achieve pollution control objectives, are wasteful. A number of economics-based incentives, such as emissions trading, encourage pollution-control activities by companies and individuals and reduce the overall costs of achieving environmental protection targets. Still, the measurement of cost/benefit (or value) to a component of the environment or to human welfare will always involve scientific factors such as aesthetic, cultural and humane values that are impossible to quantify except through the democratic process.

Global Climate Change - No environmental issue is more contentious or fraught with scientific uncertainty than the potential effect of carbon dioxide emissions on global climate change (Suplee, 1998). The debate is not about the validity of the "greenhouse effect," a real phenomena that can be proven scientifically (Charlson and Heintzenberg, 1994). It can be demonstrated that atmospheric water vapor and carbon dioxide, results in the retention of heat on the earth's surface. Indeed, the earth would be some thirty degrees Centigrade colder than normal if the greenhouse effect didn't occur (Sadler, 1996). Rather, the debate centers on whether carbon dioxide and other emissions (excluding water vapor) generated from human activities and other natural processes, which account for slightly less than 4% of all greenhouse gases in the atmosphere (Figure 4-2), are responsible for recent and future changes in global temperature.

Figure 4-2
The Most Abundant Greenhouse Gases

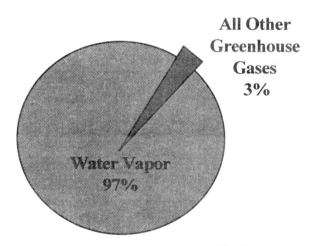

Source: World Resources Institute, 1996

Water vapor is the predominant greenhouse gas. Therefore, any human or natural activities that change the flux of water from the surface of the earth with the atmosphere will have the greatest effect on the temperature of the earth. It is likely that rapid global development will accelerate changes in the topography of the earth as a result of deforestation, fires, mining, drainage of swaps, agriculture, buildings and roads. Models will be needed to assess the effect of these changes on atmospheric concentrations of water and the resulting effect on global climate change.

During the past 18 years, satellites have been used to accurately monitor the temperature of the earth (NASA, 1998). These measurements have shown that no significant change in the average temperature of the globe has occurred during this period. It is obvious that much more work is needed to assess historical temperature data and to continue monitoring climatological changes by satellite.

A report of the Intergovernmental Panel on Climate Change (IPCC) concluded that human activity has had a "discernible" effect on global climate and postulated that there may be more pronounced effects in the future. Many scientists believe that it will be at least a decade before there is sufficient data to enable computer models to establish any relationships between global warming and human activity. This is because there are a multitude of factors that may influence average yearly global temperatures such as 1) ambient concentrations of carbon dioxide and other greenhouse gases, 2) solar (sunspot) activity, 3) changes in atmospheric water vapor concentrations caused by forest loss, buildings, desertification and paved surfaces, 4) changes in stratospheric ozone, 5) increases in atmospheric particles and 6) numerous other variables. Current climate models need to be expanded to take these multitude of factors into account.

Even though scientists have only begun to be able to model and understand the effect of the multitude of variables that may affect climate, the public has been led to believe by the media that the globe is facing 1) scorching summers, 2) dying forests, 3) melting of the polar ice-caps, 4) flooding of coastal cities and low-lying islands 5) increases in tropical diseases and 6) severe storms.

The key point is that a scientific basis for the quantitative effect of greenhouse gases on climate change has not been well developed and much more research is needed.   Ford supports international symposia and workshops which bring together the top scientists in the world to discuss research results and recommend scientific studies that will be needed to better define the potential effect of human activity on global climate change (Dahlem Conference, 1995).

Exposure to Respirable Particles - The U.S. Environmental Protection agency recently promulgated new ambient air standards to help reduce the concentrations of fine particles and ozone. Cities and states have until 2002 to develop control strategies and then up to 10 years more to meet the standards.

Several epidemiological studies have suggested that fine particles (<10 microns in diameter) have detrimental effects on human lung function, especially upon elderly people with cardiac and respiratory disease and young children with serious asthma. The proposed new standards are based upon recent epidemiological studies that have determined the relationships between increased daily mortality and an increase in fine particles.

While the data is compelling, there definitely is not a consensus among scientists as to the sufficiency of these studies as a basis for rule making. Research is proceeding in an attempt to determine the mechanism(s) responsible for potential health effects from exposure to respirable particles.

<u>Life-Cycle Assessment</u> - The three big E's of transportation are **Economics**, **Efficiency** and the **Environment**. Life-cycle assessment has become a key methodology used by scientists and engineers to help evaluate the total of all possible effects of a process or commercial product in each "**E**" category.

Ford has developed a "design for the environment" process that systematically considers environmental improvement opportunities in each part of the process used for the production of a vehicle including 1) raw material selection, 2) the manufacturing and assembly process, 3) vehicle use and 4) recycling of the vehicle at the end of its useful life. This process is more commonly referred to as "life-cycle assessment."

Life-cycle assessment is also a useful tool for emerging countries. As outlined earlier in this paper, energy security has become a top priority for China's energy policy-makers. In order to support sustainable development and energy independence, it is strategically important for China and other nations to utilize their domestic energy resources efficiently, with a minimum of environmental impact and at the lowest cost possible.

In support of this strategy, a collaborative study was sponsored by Ford and China's State Science and Technology Commission, and carried out by more than forty technical experts from eight government, industry and academic organizations in the U.S. and China, to develop and utilize life-cycle assessment (LCA) models to compare the environmental, energy and economic impacts of utilizing coal as an automotive fuel which 1) provides the lowest total cost to the customer (**Economics**), 2) minimizes the total effect on the local and global **Environments** and 3) maximizes energy **Efficiency**. This program is referred to as the "Triple E" study (Ford Motor Company, July, 1997; Kreucher, 1998).

Coal may be converted to a number of vehicle fuels such as methane, methanol, gasoline and diesel. In addition, the electrical energy generated from coal-fired power plants can be utilized to power electric vehicles. There are a number of processes that have been developed for the production of these alternative fuels and each process has advantages and disadvantages in terms of cost, energy efficiency and emissions. The "Triple E" study selected the most feasible coal conversion processes for input to the LCA models. Other options included in this study were the use of coal bed gas and coke oven gas to produce methanol.

A model was developed that provided a total accounting for each step in the life cycle including 1) the mining and transportation of coal, 2) the conversion of coal to fuel, 3) fuel distribution, 4) all materials and manufacturing processes used to produce a vehicle, 5) vehicle operation over the life of the vehicle and finally 6) recycling at the end of the vehicle's useful life (Figure 4-3)

Figure 4-3
Life Cycle Assessment for the Conversion of
China's Coal Resources to Transportation Fuels

The LCA results for all fuels were compared to gasoline, derived from petroleum, as a base-line case. Gasoline or diesel fuels derived from petroleum are the lowest cost options at this time and possibly for the foreseeable future. However, it is expected that all of the coal-derived fuels and the utilization of these alternative fuels for transportation will be changing in the direction of lower cost, more energy efficiency and lower emissions. Moreover, these other approaches could provide additional benefits related to energy security, balance of payments, etc. Therefore, it is important to understand their relative cost implications.

No single coal-derived fuel was found to be the best option in terms of all three aspects of energy, the environment and economics. However, the computer model and results generated from this study will allow policy makers to better understand how to balance energy, economic and environmental policies related to the most effective use of China's coal resources.

This study concludes that electric vehicles, even with current advanced nickel metal hydride batteries, are not presently attractive because of high vehicle cost, low energy efficiency and high life-cycle emissions of sulfur dioxide and particulate pollutants. Major advances in battery technologies will be needed, power plant efficiencies and emission control technologies will need to be improved and vehicle costs will need to be significantly reduced to make electric vehicles a viable option for the future.

In the near-term, the production of methanol from coal and coke oven gas, supplemented by the co-production of methanol from the more than 1,000 ammonia plants in China can be attractive on a regional basis. Although the cost of methanol fuel will be higher than that of gasoline or diesel, it is the most attractive of the coal-based options at this time. Methanol can be made more attractive to the consumer through appropriate subsidies or taxes.

This collaborative study was the first use of the life-cycle approach to study simultaneously the economic, energy and environmental tradeoffs associated with the use of a country's major resource. The LCA model developed from this collaborative study will help China's central and provincial governments make intelligent decisions regarding the sustainable use of fuel and energy resources.

China has been moving toward a free-market approach in which a proper balance between legislative, administrative regulations and economic incentives are being considered in developing China's transportation policies. These approaches have been developed by drawing on the experiences from other market economies.

Ye Qing, Vice Chairman of China's State Planning Commission, in his assessment of this approach, said that "the government policy makers must make difficult choices among priorities such as competing social, economic and environmental demands and severe fiscal limitations. A coherent and rational policy should include a proper mix of taxes and subsidies to achieve the environmental, energy and transportation goals that China will need to maintain sustainable growth."

Such life-cycle assessment models can be expanded in the future and used to carry out region specific studies on a global scale. The future challenge will be how to develop models that will help policy makers understand the trade-off's between energy, the environment and economics.

We expect that computer-based tools will become available in 15-20 years that can help provide government policy makers with a multi-discipline systems approach to urban areas. Such tools, while limited by present knowledge, nonetheless can help provide countries such as China with an efficient urban and inter-urban transportation system while achieving a goal of optimizing energy use and reducing the emission of pollutants and greenhouse gases. Indeed, these same tools will be especially useful to the mature economies as they build more sustainable societies.

Chapter 5
The "Global Revolution" and the Automobile

The beginning of 21ˢᵗ century will signal the dawn of the "Knowledge or Information Era" as an integral component of the "Global Revolution." As described in the preceding chapters, the primary drivers for the development of new automotive technologies during this period will be energy efficiency, the environment, economics and safety.

In order to effectively meet these challenges, transportation companies will need to develop: 1) alternative fuel vehicles,   2) high energy efficiency vehicles, 3) low emission vehicles, 4) small personal use vehicles, 5) eletricycles and 6) intelligent transportation systems (ITS) (Table 5-1).

Table 5-1:   New Transportation Technologies for the 21ˢᵗ Century

Last 100 Years	Next 100 Years - Challenges
Petroleum Fuel Vehicles	Alternative Fuel Vehicles
Low Energy Efficiency Vehicles	High Energy Efficiency Vehicles
Vehicle Emissions	Low Emission Vehicles
Large Vehicles	Small Personal Use Vehicles
Motorcycles	Electricycles
Conventional Transportation Systems	Intelligent Transportation Systems (ITS)

During the past 100 years, the primary source of fuel for vehicles has been petroleum since it has been a widely available commodity that is relatively easy to produce and distribute at a low cost. History has proven that such commodities will be available at a stable price until that resource is nearly exhausted, unless a global crisis occurs and/or government mandates are imposed. Once that resource approaches depletion, the price will rise quickly. There are many recent examples of this economic model such as the depletion of Teak, Mahogany and Redwood forests; fish stocks such as King Salmon; and even drinking water in many parts of the world.

As outlined in Chapter 3, it is estimated that the world's easily extractable petroleum resources will last approximately 40 years. Therefore, it is anticipated that petroleum will remain at a relatively stable price until sometime around 2030. The price of petroleum derived fuels will increase and alternative fuels such as natural gas and coal-derived fuels will become more cost competitive.

This eventual depletion of the world's fossil fuel resources and concerns about global climate change have resulted in the establishment of major research and development efforts by the global automotive industry to develop alternative powertrain technologies for increasing fuel economy without increasing cost and emissions, or exhausting natural resources.

Alternative Fuel and Energy Resources

Table 5-2 summarizes the primary alternative fuel and energy resources that are predicted for vehicles of the future. Estimates are given for the time periods in which these resources could represent more than 10% of the total global energy used by vehicles.

Table 5-2: Primary[1] Alternative Fuel and Electric Power Resources for Vehicles of the Future

TIME PERIOD	ALTERNATIVE FUELS	ELECTRIC POWER
2030-2100	NATURAL GAS (NG) AND NG DERIVED	NG COMBUSTION
2030-2250	COAL-DERIVED	COAL COMBUSTION
2030 and Beyond		NUCLEAR POWER
2050 and Beyond[2]	RENEWABLES	HYDRO, WIND, SOLAR, GEOTHERMAL

[1] Primary defined as representing more than 10% of the total global energy requirements for vehicle transportation

[2] All renewable and alternative electric power sources combined projected to represent more than 10% of the total global energy requirements for vehicle transportation

The world's natural gas resources will last about 60 years at the current rates of use. In addition, it is estimated that there is probably an additional 50% of today's known resources that have not been discovered. Because natural gas is an abundant, environmentally friendly, efficient, low-cost and flexible fuel, we predict that it could become a major (>10% of total global vehicle energy use) alternative source of transportation fuel sometime around 2030.

Natural gas can be used in a variety of ways to power a vehicle. Such options include: 1) a compressed natural gas (CNG) vehicle, 2) a bi-fuel system with the option to switch to a liquid fuel, 3) an electric or hybrid-electric vehicle after the natural gas is converted to electricity in a thermal-electric power plant, or 4) a natural gas derived fuel such as dimethylether, dimethoxyether, gasoline or diesel.

Although, most of the world's global automotive companies have produced many commercial and prototype versions of alternative fueled vehicles, the infrastructure for alternative fuels is basically non-existent. The fact that the infrastructure for gasoline fuels is valued at about $200 billion (not including land and building costs) provides a good perspective on how expensive it will be to develop an infrastructure for alternative fuels and energy sources. As a result, it is expected that the first uses of CNG will be limited to large vehicles such as buses and trucks and vehicle fleets such as delivery vans and taxis.

Ford has been the most proactive automotive company in the advancement of alternative fueled vehicle (AFV's) technologies. Ford equipped AFV's account for more than 90% of all AFV's sold in North America in 1997. Ford is producing six natural gas vehicles in the 1998 model year. Over 250,000 flexible-fuel minivans and pickup trucks will be built in a four-year period that can run on various mixtures of ethanol and gasoline. It is expected that other vehicle manufacturers will become more proactive and build alternative fueled vehicles to help encourage alternative fuel infrastructure development.

Alternative fueled vehicles that utilize natural gas, propane, alcohols and ethers will gradually begin to replace gasoline fueled vehicles. Natural gas vehicles will gradually replace gasoline fueled vehicles as natural gas fueling stations become more available.

The development of technologies to produce synfuels from natural gas and coal have been actively pursued since the early 1920's. However, these fuels have not had a major impact during the 20th century since they have not been cost competitive with fuels derived from petroleum. Synfuels will become implemented in Asia before other parts of the world as Asia's oil resources become depleted and more costly. The most promising coal-derived and methane derived alternative fuels are dimethylether and dimethoxyether. Dimethylether has nearly the same vapor pressure as propane and it can be used as an alternative to LPG fueled vehicles. Dimethoxyether can use the same fuel infrastructure as gasoline. The use of these fuels in diesel engines can result in a significant reduction in particle, hydrocarbon, NOx and CO emissions.

A dimethoxyether fueled diesel engine in a hybrid electric vehicle could exemplify the low-emission, high-energy efficiency powertrain of the future. This may be the only fuel, combined with advanced diesel engine and emissions control technologies, that will be able to meet the goal of 10 mg of particle emissions per mile and the U.S. automotive industries Progress for a New Generation of Vehicles (PNGV) goal of 80 miles per gallon.

The cost of producing electric power from a natural gas plant is about 40% lower than that from coal-fired plants. This lower cost is the result of the lower capital expenditures for a natural gas plant and the energy efficiency of power production of nearly 60% for natural gas compared to 35-40% for coal.

Gasoline and diesel fuels can be derived from coal but the equivalent cost per barrel of oil is currently at about $30 (Kreucher, 1998). Coal-derived fuels will become more attractive as petroleum supplies dwindle and become more expensive. A major challenge for the production of coal-derived fuels will be to reduce the carbon dioxide emissions. If this problem can be solved and the cost of production reduced, we predict that coal-derived fuels could become a major alternative energy source sometime in the 2030-2040 time frame.

The use of renewable energy resources for vehicles will become more attractive as new technologies increase energy efficiency and decrease costs. Alternative electric power technologies include hydroelectric, wind driven turbines, solar power and geothermal sources. Alternative renewable fuels include ethanol, methanol, methane, hydrogen, diesel, and others.

A number of new methods are being developed for the energy efficient and cost-effective conversion of waste organic materials into ethanol, methanol and methane. Renewable fuels such as ethanol from biomass will become more practical. Cellulosic ethanol is a promising biofuel made from agricultural, forestry and domestic wastes that is projected to cost (U.S. Dept. of Energy, 1995) under $0.70 per gallon by 2005. Countries, such as most of those in SE Asia, tropical South America and Africa, with very long growing seasons and ample rainfall, could produce substantial quantities of ethanol from their agricultural and forest waste. However, total production of these renewable fuels is not likely to represent a large fraction of the total demand by transportation.

As the cost of electric and hybrid electric vehicles become more cost competitive and of comparable reliability to current gasoline and diesel-powered vehicles, electricity will become an important source of energy for vehicles. The flexibility, accessibility and affordability of electric power will facilitate the acceptance of electric vehicles in the 21st century.

Electric power has the advantage that it can be generated from a multitude of sources that can be easily fed into the world's rapidly expanding electrical grid system. By the end of the 21st century it is predicted that more than 95% of the world's population will have easy access to an affordable supply or electricity.

A move from traditional to alternative engine and powertrain technologies will represent a quantum step for the $1 trillion/year global automotive industry. If these new technologies become cost competitive and reliable within the next 20-30 years, they will require tremendous investments in new tooling, ultimately render obsolete most of the world's 600 million cars and trucks, and change every gas station on the globe. Automobile companies will become known as transportation companies as they broaden their scope, the oil industry will probably evolve into a more diversified energy industry, and gas stations will become energy stations.

High Energy Efficiency Vehicles

Although electric power has been abundantly available in the mature markets, and electric vehicle technology has been available for nearly 100 years, this technology has not become practical on an economic, environmental or energy standpoint.

At the present time, barring any breakthrough in electric power storage technologies, the hybrid electric vehicle (HEV) is a much more attractive future concept than the electric vehicle. This is not a new concept, but one developed by H. Piper in the early 1900's to help boost the power of internal combustion engines. Recent interest in this technology has been developed as a method to conserve energy and reduce emissions.

The energy efficient vehicles of the future could incorporate a hybrid-electric propulsion system including: 1) a solid-state energy storage system which may be electrochemical (battery) or thermal; 2) a fuel tank to store liquid fuel for an internal combustion engine; 3) an internal combustion engine, heat engine or fuel cell stack and 4-6) an energy transfer system consisting of power electronics, a high-efficiency automatic transmission and a direct-drive electric motor (Figure 5-1).

Figure 5-1
High Energy Efficiency Vehicles of the Future
Hybrid Propulsion

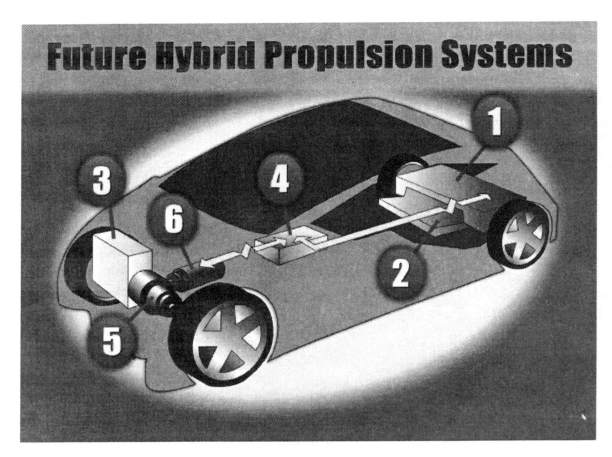

Energy Storage Systems
1) Solid-State Energy Storage (electrochemical or thermal)
2) Fuel-Tank (liquid or gaseous fuels)

Energy Conversion Systems
3) Internal Combustion Engine, Heat Engine or Fuel Cell Stack

Energy Transfer Systems
4) Power Electronics
5) High-Efficiency Automatic Transmission
6) Direct-Drive Electric Motor

Energy Storage Systems - There are a number of options available for the storage of energy in a vehicle (Figure 5-2). Such options include the conventional liquid fuels such as gasoline or diesel; and alternative chemical fuels such as ethanol, methanol, liquefied petroleum gas (LPG), liquefied natural gas (LNG), liquefied hydrogen (LH2), compressed natural gas (CNG), compressed hydrogen (CH2) and dimethyl ether. Electrochemical storage systems include lead acid, cadmium nickel, lithium polymer and lithium metal hydride batteries.

Figure 5-2
Energy Densities of Alternative Fuels Relative to

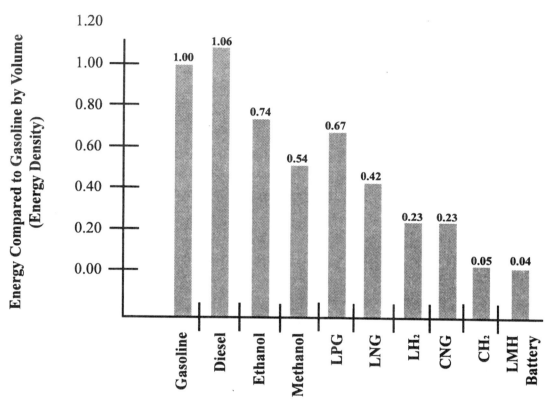

The liquid alternative fuels ethanol, methanol, LPG and LNG have energy densities (energy content per unit mass) that are from 42% to 74% of that of gasoline. Compressed natural gas (CNG) needs to be stored at 3500 psi pressure to represent 23% of the energy density of gasoline. Hydrogen has to be liquefied at nearly -253 C to provide the same energy density as CNG. Although hydrogen is a clean, renewable fuel, it is not possible to store enough hydrogen, even at 3500 psi, to provide much range for a vehicle.

The problem with liquid and gaseous hydrocarbon fuels is their poor thermal efficiency when used in current internal combustion engines (~25%). However, it is expected that IC engines with thermal efficiencies of up to 40% could become available in the next 20 years.

It is doubtful if electrochemical batteries will exceed an energy storage density that is greater than 5% of gasoline (Figure 5-2). However, it is theoretically possible to store heat with an energy density that is at least equivalent or possibly twice that of gasoline. The future technical challenge will be to develop low-cost methods to efficiently convert that stored heat to kinetic energy.

Energy Conversion/Energy Transfer Systems - Currently, the most popular HEV concept is a hybrid of the electric car with a small internal-combustion (IC) engine and an electric generator on board to charge the batteries (Figure 5-1). The power of the IC engine is approximately half that of a conventional automobile. This engine can run at a speed that optimizes energy efficiency and reduces engine emissions. The batteries may be charged continuously or only when they become depleted to some level.

Other hybrid systems under development include heat engines with electric motor drives and fuel cells with electric motor drives (Table 5-3). One or more of these options has a high probability of becoming an important powertrain system of the future. However, it is not expected that the combined economic, environmental and energy advantages of these powertrains will be realized until after 2030. We predict that vehicles of this type could reach a market penetration of 10% or greater at this time. One reason for this long lag time is that automakers will continue to improve gas and diesel burning engines to reduce emissions and improve fuel economy.

Table 5-3: Energy Conversion Systems for Vehicles of the Future

TIME PERIOD	IC ENGINES[1]	HYBRID ELECTRIC SYSTEMS[2]
2030 AND BEYOND	IC ENGINE	IC ENGINE/ELECTRIC FUEL CELLS/ELECTRIC HEAT ENGINE/ELECTRIC

[1]Includes spark ignition and compression ignition engines operating on conventional and alternative fuels
[2]All powertrains listed as representing more than 10% of the total global fleet of vehicles.

The fuel cell is another example of an old technology (invented in 1839) which has garnered renewed interest for automotive applications. Fuel cells may become a major vehicle powerplant of the future if some major obstacles can be overcome. The advantages of using a fuel cell include higher energy efficiency, lower emissions and the potential ability to operate on hydrogen (H_2), or on fuels from which hydrogen can be produced such as ethanol, methanol, and natural gas. A methanol fuel cell can extract about 80% of the energy in a gallon of methanol compared to about 25% for a gasoline engine. The emissions of regulated pollutants from a fuel cell vehicle using methanol would be more than 10 times below that of current low-emission, gasoline vehicles.

Much of the development work on methanol fuel cells has been carried out by Daimler-Benz AG. Ford formed an alliance with Daimler-Benz AG and Ballard Power Systems, Inc. of Canada in 1997 to develop methanol fuel cell vehicles that could result in the production of up to 100,000 cars a year by 2004.

The biggest issues facing the fuel cell will be cost and long-term reliability. Current fuel-cell powertrains cost about $30,000, compared to a comparable 4-cylinder IC engine and transmission costing about $3,000. Future fuel cells could be developed that utilize rare-earth oxide materials combined with small amounts of precious metals to reduce costs.

Ford has developed an electric hybrid concept vehicle, called the P2000, that delivers over 63 mpg in a 5-seat midsize vehicle. This vehicle weight is about 40% less than comparable vehicles, without compromising safety, performance, durability, comfort and affordability. Other major auto manufacturers are considering the mass production of hybrid electric vehicles in the 2002-2010 time frame.

Another option that is being pursued is the fuel reformer/fuel cell/electric motor hybrid. The reformer is an on-board chemical reactor that produces a fuel, such as hydrogen, which is used by the fuel cell to produce electricity. This system is attractive because current fuels such as gasoline can be reformed into simpler fuels. However, oxidation reformers of contemporary designs have limited lifetimes since oxidation products and polymeric materials can accumulate and they often involve fragile or easily contaminated catalysts. These problems must be overcome if this hybrid technology is to become competitive.

Low-Emission Vehicles

A great deal of progress has been made during the past twenty years toward the development of control technologies to reduce vehicle emissions to low- and ultra-low levels. The future challenge will be to develop cost-effective and reliable technologies for the reduction of diesel particle and NOx emissions to levels that are comparable to those from gasoline vehicles. The development and implementation of technologies for the use of low-emission diesel fuels such as ethers will probably be necessary to meet these goals (U.S. Dept. of Energy, 1998).

The future global implementation of low emission vehicles will be constrained by the general unavailability of high quality gasoline and diesel fuels, an insufficient local source of emission control components, and the lack of an infrastructure for vehicle inspection and maintenance in most emerging markets.

There is also the issue of affordability and effectiveness. The emerging markets have a multitude of environmental problems to solve. Their current priority is to make clean water and waste disposal widely available. Another top priority is to reduce emissions from the wide-spread use of coal.

Even if these hurdles were overcome by the 2003-2005 time frame, it would still require another 10-15 years to replace most of the in-use vehicles. Therefore, it is projected that it will take until about 2020 to replace all in-use vehicles with low emission vehicles and that economics, not environmental factors, will probably prevail as the force guiding the industry.

Personal Use Vehicles

Small, light commuter type vehicles have been available for over 50 years. These "minicars" typically weighed 300-650 kg and were designed for 1-3 occupants. An assessment of the European experience with minicars can provide important insights into the potential popularity of such a vehicle for the 21st century.

Most of the minicars developed for the European market during the past five decades have become extinct. The primary reason for the failure of this market segment is that families desire good overall value in terms of transporting maximum payload in comfort with minimum vehicle expenditure, especially if only one car is available per family. The minicar has not offered this value.

Many of the emerging families of the 21st century will have limited purchasing power. The key vehicle attributes that such families will want are reliability, safety, good value, multi-purpose configurations, comfort and low operating costs. If such a vehicle can be produced at a cost equivalent to $7,000-8,000 U.S., the market potential will be enormous (Schuetzle, 1996).

It is expected that one or more manufacturers, could produce this "Model T equivalent" of the 21st century within the next 10-15 years. The introduction of such vehicles will help the automotive industry meet the energy and environmental challenges of the future.

Electricycles

Although there are 600 million vehicles worldwide, about one billion people still rely on 2-wheel and 3-wheel vehicles as their primary mode of transportation (World Resources Institute, 1996). During the next 100 years, many of these people will progress from human powered to motorized 2-wheeled vehicles, especially in urban areas.

The major problems associated with the use of internal combustion (IC) engines for motorcycles are their high level of emissions and noise. Unless there are substantial solutions to these environmental problems, the city governments of the world's major urban areas could establish policies that require the use of electric 2- and 3-wheeled vehicles or "electricycles" to replace these motorcycles. These "electricycles" also have some advantages for rural areas in which fuel may be scarce and/or of poor quality. They may be charged using renewable energy sources such as wind-driven generators and solar cells.

There are a number of companies that are developing some very innovative, low-cost and robust "electricycles." One of these companies, "ZAP" (stands for Zero Air Pollution), has a number of electric and pedal power hybrids as well as electric scooters and motorcycles. Some novel features of these electricycles include easy battery removal to facilitate charging at the owners place of residence or for exchange at a convenience store (Figure 5-3), an inexpensive solar panel charging system, and accessories for converting nearly any bicycle to electric power. Several Asian cities such as Shanghai, Bangkok and Guangzhou are considering the establishment of policies that will favor the replacement of gasoline powered 2-wheeled vehicles to electricycles.

In Chapter I, we described how the members of the New York Gentleman's Cycle Club promptly traded in their bicycles for new Model T's during the early 1910's. Although we don't expect urban dwellers of the 21st century to trade in their personal vehicles for electricycles, we do expect that these vehicles will become more desirable as an additional, alternative form of transportation in congested cities. The urban commuters of the future may carry an electricycle in the trunk for travel in the core urban areas in which vehicle use will be limited.

58

Figure 5-3
ZAP Electric Motor Scooter with Removable Battery Packs

Source: ZAP, 1997

Intelligent Transportation Systems

Since Henry Ford's time there has been a whole sequence of technological innovations, but little has changed with respect to the interaction of the vehicle with the highway. However, that outlook will change significantly in the future. The global implementation of intelligent transportation systems (ITS), including intelligent vehicle systems (IVS) and the promulgation of futuristic urban transportation policies will significantly increase travel efficiency, enhance personal productivity, reduce emissions and conserve energy.

ITS is an approach consisting of many possible new highway and vehicle subsystems that use advanced computer and electronic technologies for vehicle and transportation improvements (Figure 5-4). The key enablers for the widespread use of ITS technologies will be low cost, high-speed computers, globally positioned satellite systems, a vast global computer infrastructure, information technology and knowledge management.

Figure 5-4
Intelligent Transportation System
Concepts of the Future

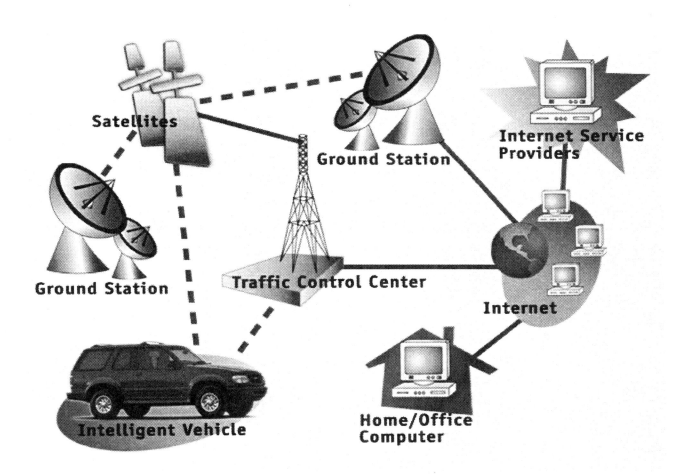

Personal success in the future will be dependent upon how effectively people harvest the most useful information from the increasingly vast sea of data, assemble it to make decisions, develop new technologies, and market products and services. The vehicle should become an extension of the office and home in this capability.

Several recent studies have demonstrated that traffic congestion can have a significant negative effect on the productivity and environment of an urban area. It has been estimated that Bangkok, Thailand could increase its gross domestic product output by as much as 30% by eliminating traffic gridlock. A study in Koln concluded that as much as 60% of the congestion in that city is the result of people searching for a place to park.

During the next century, the development of Intelligent Vehicle Systems that can access the information superhighway will be a high priority for automotive manufacturers. IVS systems will allow drivers to obtain updates on traffic conditions, to receive help with directions, and to send and receive information using voice and retinal and possibly tactual and neural interfaces with their vehicles.

It will be more important in the future that vehicles have the capability to connect the driver and passengers effectively to the rest of the world without affecting safety. "Intelligent Vehicles" will become standard equipment. These vehicles will come with satellite links to access Internet, e-mail and voice mail, as well as cable and broadcast channels (Ward, 1997). By 2100 nearly every vehicle on the globe will be interconnected by inexpensive, rapid (transfer rates up to 50 mbytes/sec.), wireless communication systems.

Network information systems will become available that provide a real-time interactive routing system with directions to a destination in the shortest time and distance possible. Other systems will allow a driver to find the closest available parking spot and the businesses that provide the goods they wish to purchase at the lowest price in a specified area.

Charles Trimble, CEO of Trimble Navigation, has predicted that within the next 100 years "GPS enhanced positioning guidance systems with geocode link to the internet will not only make getting lost or turned around nearly impossible, but will provide the basis for efficient voice and data communication, with people in cars, homes and offices, around the world. Furthermore, this technology will form the basis for autopilot drive and navigation without the need for specially constructed high speed corridors" (Trimble, 1998).

GPS systems in coordination with vehicle vision systems and vehicle location systems will allow vehicles to be detected globally with an accuracy of five cm or less. This information will be processed and transmitted from regional and global ITS coordinating centers. You will not only know exactly where your vehicle is at any particular time, but you will be able to quickly determine the location of any other vehicle in the world. Other features will include navigation systems that provide directions and diagnostic systems to detect problems such as malfunctioning emission control systems.

The "Intelligent Vehicle" will become a complimentary partner of the driver. The vehicle will have a "real-time" awarness of environmental conditions, the movement of objects within the range of the vehicle and potentially dangerous situations. The vehicle will assist the driver by verbal warnings and evasive actions to avoid danger.

These technologies could make possible the global deployment of "Auto-Pilot Corridors" in which vehicles can enter and travel without driver intervention. Vehicles would communicate and synchronize their positions with other vehicles on the road. The system will adjust velocities for road conditions and other environmental factors. Such corridors may require high bandwidth wireless roadside cameras and road imbedded sensors to help insure a high degree of reliability and safety. Such systems would help prevent collisions, reduce congestion and cut emissions. Although the implementation of ITS will lead to environmental and energy improvements, the principle drivers for these technologies will be primarily the improvement of transportation efficiency and safety.

The California Department of Transportation (CalTrans) recognized in the mid-1980's that it would be necessary to rely on advanced technologies to satisfy the freeway capacity needs of the Los Angeles region, early in the new century. Consequently, they formed a partnership with the major research universities in the state, led by the University of California at Berkeley, to establish the California PATH Program to conceptualize, develop and demonstrate these technologies.

The centerpiece of the California PATH development is the automated highway system, which has been envisioned by futurists since the 1930's, but was finally demonstrated publicly in 1997 by PATH and its partners in the National Automated Highway System Consortia. PATH Deputy Director, Steven Shladover, points out that "the automated highway system represents a new level of integration between vehicle and infrastructure technologies, permitting an unprecedented combination of improvements in highway capacity, safety, efficiency and environmental quality" (Shladover, 1998).

An examination of "auto-pilot" systems for aircraft can help predict the long-term potential for automated highway systems. Although auto pilot systems are used at high altitudes and during long flights, the pilot and co-pilot are constantly monitoring weather conditions and other aircraft. Sophisticated electronic devices are used to help guide the plane during landing, but the pilot retains full control of the aircraft.

The ultimate success of "auto-pilot" technologies for vehicles will be dependent upon infrastructure costs and most importantly the safety of these systems. Even though complex strategies for nearly 100% "fail-safe" operation can be programmed into such systems, we believe that they will never be reliable enough to operate without driver intervention. This is because the potential for danger is much greater for a vehicle on the road than an aircraft at 30,000 ft. A vehicle on the road has to deal with orders of magnitude more moving and non-moving hazards than a vehicle in the air.

Future of Global Mobility

An assessment of the future modes of transport are critical for planning infrastructure and for assessing the consequences of mobility. A number of studies have been carried out (Schafer and Victor, 1997) to develop scenarios for the future volume of passenger travel, as well as the relative prevalence of different forms of transportation through the year 2050. It has been found that people devote a predictable fraction of their income to transportation. This fraction is 3-5% in emerging economies and 10-15% in mature economies. They also found that people devote a constant fraction of their daily time to travel. This fraction typically averages between 1.0-1.5 hours/person/day for people living in a wide variety of economic, social and geographic settings. Therefore, it is projected that the traveler of the future will select faster modes of transport to cover more distance within the constraints of their travel time budget.

Historical data shows that, throughout the world, personal income and traffic volume grow in tandem. As average income increases, the annual distance traveled per capita rises by roughly the same proportion. Therefore, it is projected that by 2050, high-speed transport should account for about 40% of all passenger miles traveled, compared to about 15% today. Automobiles will supply about 35% of the total passenger miles traveled, compared to about 48% today.

Although the demand for air travel or high-speed rail will increase substantially by 2050, the global high speed travel budget per person is projected to be 12 minutes per day. Most travel time for people living in emerging countries in 2050 will still be by walking and 2 or 3 wheeled, low-speed vehicles (less than 25 miles/hour maximum speed). Therefore, it is projected that low-cost, environmentally friendly vehicles such as electricycles could become an important form of transportation for up to one billion people by the end of the 21st century.

Chapter 6
Conclusions

This paper has not attempted to prophesize the future, but it does outline the probable scope of some future scenarios. New automotive technologies will not leap from the laboratory to the mass-market overnight. The long-term nature of research and the high potential for failure, suggest that many options should be pursued at once. Some new technologies can best be tested in selected markets where there are special needs and resources, or other advantages. Examples of such opportunities include low-cost, environmentally friendly, robust hybrid-electric vehicles; intelligent transportation systems; and efficient, personal use vehicles for high-density urban areas. Costs will fall rapidly as these and other new technologies become more widely accepted.

There is also the possibility that entirely new technological innovations will come forward to change the automotive industry in fundamental ways. For example, if new technologies can be developed to efficiently convert coal into environmentally acceptable fuels, a carbon-based energy economy may endure for a much longer time.

During the past century, we have seen the automotive industry grow in parallel with the aircraft and aerospace industries. At moments during the development of these industries in the 20th century, it appeared that the two might merge into one. Such examples have included Ford's production of aircraft from the 1920's through the 1940's, the formation of Ford Aerospace in the 1960's, and the acquisition of Hughes aircraft by GM in the 1980's. One example of the collaboration with Ford, General Motors and NASA resulted in the design and operation of the first vehicle on the Moon (Figure 6-1).

Figure 6-1
Model of the GM/Ford/NASA Lunar Rover (1969)

Both industries and the global environment have benefited by this close association. One key example of this benefit has been the vehicle applications of on-board microprocessors developed as a major technology effort in the moon program.

Although Ford and GM have reduced their efforts in the aerospace industry, we anticipate that the aerospace and automotive industries will again begin to work more closely together and in some cases establish significant business relationships. It is predicted that these new alliances will facilitate the development and production of vehicles for operation on the Moon, Mars and possibly other celestial bodies in the later part of the 21st century.

Although there are many environmental problems facing the emerging markets at this time, these markets appear to be paying more attention to environmental problems than Europe or America did at similar levels of adjusted per capita income. For example, China has adopted sophisticated environmental regulations and they have established regional environmental protection offices throughout the country.

Even though these regulations are not strictly enforced at this time, China's State Environmental Protection Administration (SEPA) will gradually tighten laws in the coming years. A key imperative of the Chinese government has been to raise the awareness of policy makers and the public. This will ensure that local economic and social development policies are friendlier to the environment. Other emerging countries are adopting similar approaches and policies. Mr. Xie Zhenhua, Minister of China's SEPA, believes that "There are no national boundaries in handling environmental issues. The Earth is big, yet it is only a global village whose residents should co-operate closely to preserve the surroundings."

The rate of environmental degradation in China and other emerging nations is expected to worsen until the 2005-2010 time frame, and then gradually improve. Even though there are substantial efforts in the area of environmental protection, it is anticipated that a high level of environmental quality for all nations will not be achieved and sustained until about 2020-2030 time period.

One of the most encouraging examples for the future direction of the emerging nations has been Japan. Japan was perhaps the most polluted country in the world in the early 1970's. It transformed itself into one of the least polluted and most energy conscious nations within 20 years.

The effect of the automobile industry's efforts to improve energy efficiency will help boost the global economy by spurring innovation. Those companies that produce cost-effective, environmentally friendly and energy efficient vehicles, will maintain a long-term market advantage.

A move to alternative vehicles will require tremendous investments. Mass production of a single advanced-technology model requires several billion dollars in investment. A bet on the wrong technology could leave a company at a competitive disadvantage.

The automobile industry will have many environmental challenges to face in the future. Ford Chairman Bill Ford, who is also Chairman of Ford's Environmental and Public Policy Committee, presented a keynote address at the 14th International Electric Vehicle Symposium and Exposition in Dec. 1997 which wraps up this challenge very succinctly. He commented that "for people who are concerned about the environment, this is an exciting time to be in the automobile business. There are great challenges and great opportunities. What we need are real-world, customer-driven solutions. I think we'll find that no rule or regulation can push technology to the forefront as fast as today's competitive global market."

Acknowledgments

The authors are indebted to a great number of talented people who made possible such a comprehensive perspective and prospective view on the automotive industry and the global environment. These persons include Michelle Marcoux, literature research and graphics; Jeannette Mason, word processing; John Tyman, photography and video creation; Greg Turley, photography archives and multimedia production; Chris Merlo, editing and scripting; Gary Donnley, production management; Shelly Andrews, graphics and illustration; Mike Levy, multimedia and animation; Li Feng Xu, energy research; Al Lee, Ford Motor Co. historian and writer; Chuck Risch, electric and hybrid vehicle technology: There are a number of organizations we would like to acknowledge which helped us examine the past as well as look through the window to the future including the Henry Ford Museum, the National Academy of Sciences; WQED - Pittsburgh; Miramar Productions; the Pasadena Art Center College of Design; NRG Corporation; ZAP Corporation; Trimble Navigation Corporation. The cooperation of the city governments of Beijing, Paris, Shanghai, Bangkok and Bombay are also appreciated. In addition, the authors greatly value the expert reviews of this manuscript by Dave Wagner, Chuck Risch, John Spelich, Shirley Durham and Helen Petrauskas of Ford Motor Company; Robert Schuetzle, Digitella Corporation; Mark Jones, University of California - Riverside; Mal Weiss, Massachusetts Institute of Technology; John Kennedy, Oil and Gas Journal; Charles Trimble, Trimble Navigation; Dieter Beike, Enron Oil and Gas International; and Steven Shaldover, University of California - Berkley.

References

Alliance To End Childhood Lead Poisoning, "Gasoline Lead and Blood Lead Levels in the United States," Washington, D. C. (1994).

Asia Society, "Meeting Asia's Energy Challenges in the 21st Century: Business, Political and Strategic Implications for the United States and Asia," Houston, Texas (May 1, 1998).

Brown, S. F., "The Automakers' Big-Time Bet on Fuel Cells," Fortune, 122B-122D (March 30, 1998).

Bryan, Ford R., "Beyond the Model T", Great Lakes Books, Wayne State University Press, Detroit, Michigan 48202 (1990).

Calvert, J. G., Heywood, J. B., Sawyer, R. F. and Seinfeld, J. H., "Achieving Acceptable Air Quality: Some Reflections on Controlling Vehicle Emissions," Science (July 2, 1993).

Campbell, C. J., "The Next Oil Price Shock: The World's Remaining Oil and Its Depletion," Energy Exploration and Exploitation, Vol. 13, pp. 19-44 (1995).

Carson, R., "Slient Spring," Houghton Mifflin Company, 215 Park Avenue Suth, New York, New York 1003 (1962).

Charlson, R.J., Heintzenberg, J., "Aerosol Forcing of Climate," John and Wiley & Sons Ltd., England, (1995).

China daily Newspaper (October 27, 1997).

Coates, J. F., Mahaffie, J. B., Hines A., "2025 - Scenarios of U.S. and Global Society Reshaped by Science and Technology," Oak Hill Press, Greensboro, N.C., (1974).

Commission on Life Sciences, Committee on Measuring Lead in Critical Populations and Board on Environmental Studies/Toxicology, "Measuring Lead Exposure," National Academy Press, Washington, D. C. (1993).

Corzine, R., "Oil and Gas Reserves will all Go, says Shell Chief", Financial Times (U.K.) (Oct. 14, 1997).

Environmental Research Institute of Michigan - ERIM International, Inc. (Internet Site: www.erim.org) (1998).

Ford Motor Company, "Economic, Environmental and Energy Life Cycle Assessment of Coal Conversion to Automotive Fuels in Shanxi Province and Other Coal Rich Regions of China," Dearborn, Michigan, (July, 1997).

Ginsburg, Douglas H., Abernathy, William J., "Government, Technology and The Future of The Automobile," McGraw - Hill Book Company, New York (1980).

Graedel, T. E., Allenby, B. R., "Industrial Ecology and the Automobile", Prentice Hall, New Jersey, 243 pp (1998).

Hahn, R. W., "Risks, Costs and Lives Saved," Oxford University Press (1996).

Hart, Stuart L., "Beyond Greening: Strategies for a Sustainable World," Harvard Business Review, Harvard College (1996).

Hong, Chuan-Jie, "A Review of Air Pollution and its Health Effects in China", United Nations Environmental Program, International Program on Chemical Safety, World Health Organization, 1211 Geneva 27, Switzerland (1993).

Intergovernmental Panel on Climate Change (Internet Site: www.ipcc.ch) (1997).

International Energy Agency, "World Energy Outlook," 75739 Paris Cedex 15, France, 475pp. (1998).

International Standardization Organization, ISO 14000 Certification Procedures, Geneva, Switzerland (1997).

Ivanhoe, L. F., "Future World Oil Supplies: There is a Finite Limit," World Oil, page 78 (October 1995).

Johnston, James D., "Driving America", American Enterprise Institute, Washington, D.C. (1997).

Keoleian, G. A., Kar, K., Manion, M. M., Bulkley, J. W., "Industrial Ecology of the Automobile: A Life Cycle Perspective," Society of Automotive Engineers, Inc. 400 Commonwealth Drive, Warrendale, PA 15096-0001 (1997).

Kiefer, Irene, "Energy for America," Atheneum, New York (1979).

Kreucher, W.M., Han, W., Schuetzle, D., Zhu, Q., Zhang, A., Zhao, R., Sun, B., Weiss, M.A., "Economic, Environmental and Energy Life-Cycle Assessment of Coal Conversion to Automotive Fuels in China," SAE Paper #98TLC-79, Society of Automotive Engineers, Warrensdale, PA, (1998).

Lamprecht, James L., "International Standardization Organization, Issues and Implementation Guidelines for Responsible Environmental Management", AMACOM - a division of American Management Association, New York (1997).

Lee, A., "The Inevitable Flying Car", Science and Mechanics (March, April 1983).

Lents, J. M., Kelly, W. J., "Clearing the Air in Los Angeles," Scientific American (October, 1993).

Li Peng, "China Energy Policy" (May 29, 1997).

Lomasky, L. E., "Autonomy and Automobility", in Independent Review, Independent Institute, 134 98th Ave, Oakland, CA 94603 (1997).

Scientific American, "The Future of Transportation" (Oct., 1997).

Masters, C. D., Root, D. H., Attanasi, E. D., "Resource Constraints in Petroleum Production Potential," Science, p. 146 (July 12, 1991).

McElroy, M. B., Nielsen, C. P., Lydon, P., "Energizing China: Reconciling Environmental Protection and Economic Growth," Harvard University Press (1998).

National Aeronautics and Space Administration, Earth Science Enterprise (Internet Site: www.hq.nasa.gov) (1998).

National Geographic, "Untangling the Science of Climate," National Geographic Society, Washington, D. C. (May, 1998).

National Research Council, "Fuels to Drive our Future", National Academy Press, Washington, D. C. (1990).

O'Callaghan, T. O., "Henry Ford's Airport and Other Aviation Interests - 1909-1954," Proctor Publications of Ann Arbor, Ann Arbor, Michigan (1993).

Polesetsky, Matthew, "Global Resources: Opposing Viewpoints," Greenhaven Press Inc., San Diego, California (1991).

Ronning, J. J., "The Viable Environmental Car: The Right Combination of Electrical and Combustion Energy for Transportation," Paper #971629, Society of Automotive Engineers, Warrendale, PA. (1997).

Sadler, A. E., "The Environment: Opposing Viewpoints," Greenhaven Press Inc., San Diego, California (1996).

Schafer, A., Victor, D., "The Past and Future of Global Mobility," Scientific American (Oct., 1997).

Schuetzle, D., et. al., "China Automotive Technology Workshop Recommendations," Society of Automotive Engineers, Warrendale, PA (1996).

Seiffert, U., Walzer, P., "Automobile Technology of the Future," Society of Automotive Engineers, 400 Commonwealth Drive, Warrendale, PA (1991).

Seinfeld, J. H., "Atmospheric Chemistry and Physics of Air Pollution," John Wiley and Sons, Inc., New York (1986).

Sheahan, Richard T., "Fueling the Future: An Environmental and Energy Primer," St. Martin's Press, New York (1976).

Society of Automotive Engineers, Proceedings of the 1997 SAE International Fuel and Lubricants Meeting, Warrendale, PA (1997).

Song, J., Tuan, C.-H., Yu, J.-Y., "Population Control in China: Theory and Applications," Praeger, New York, ISBN 0030698812 (1985).

Steingraber, S., "Living Downstream," Addison-Wesley Publishing Company, Inc., New York, ISBN 0201483033 (1997).

Stipp, D., "Trouble in The Air", Fortune, pp. 113-120 (Dec. 8, 1997).

Suplee, C., "Untangling the Science of Climate", National Georgraphic (May, 1998).

U.S. Bureau of the Census, International Data Base (Internet Site: www.census.gov) (1997).

U.S. Dept. of Energy, U.S.-China Energy and Environment Cooperation Initiative (October, 1997).

U.S. Department of Energy, Office of Advanced Automotive Technologies R&D Plan, "Energy-Efficient Vehicles for a Cleaner Environment," (March, 1998).

U.S. Department of Energy, International Energy Annual 1996 and Energy Information Administration - China (Internet Site: www.eia.doe.gov).

U.S. Department of Energy, "Partnership for a New Generation of Vehicles Program Plan" (Nov. 29, 1995).

U.S. Department of Energy, Scenarios of U.S. Carbon Reductions: Potential Impacts of Energy Efficiency and Low-Carbon Technologies by 2010 and Beyond 202-586-5575 (Sept. 1997).

U.S. Department of Energy, Renewable Energy Data - Renewable Energy Annual 1997 (Internet Site: www.eia.doe.gov) (1997).

U.S. Department of Energy, "Comprehensive National Energy Strategy" (Internet Site: www.eia.doe.gov/nesp/cnes.html) (1998).

U.S. Department of Energy, "Federal Energy Technology Center Home Page" (Internet Site: www.fetc.doe.gov) (1998).

Vig, Norman J., Kraft, Michael E., "Environmental Policy In The 1990's," CQ Press, A Division of Congressional Quarterly Inc., Washington, D.C. (1994).

Wagner, B., d Koronkiewicz, "ISO 14000: Environment and Business Friendly," Automotive Engineering International (April, 1998).

Walter, D., "Today Then," American and World Geographic Publishing, Helena, Montana (1992).

Ward, A. M., "How About a Trip on The Information Highway in a Network Vehicle?" Web Week (Nov. 24, 1997).

Winters, J. "Nanotanks", Discover, p. 43 (Jan., 1998).

Wolf, Sidney M., "Pollution Law Handbook: A Guide to Federal Environmental Law," Greenwood Press Inc., Connecticut (1988).

World Bank, "Clear Water, Blue Skies - China's Environment in the New Century," World Bank China 2020 Series (1997).

World Resources Institute, "Environmental Almanac," Houghton Mifflin Company, Boston and New York (1993).

World Resources Institute, The United Nations Environmental Program, The United Nations Development Program and The World Bank, "World Resources," Oxford University Press, New York (1996).

ZAP (Zero Air Pollution) Power Systems, 117 Morris Street, Sebastopol, California 95472; Tel (707) 824-4150 (1998).

Zhu Baoxia, China Daily Newspaper, Beijing, China (April 13, 1998).